INTEGRATED MODELING
OF LAND AND WATER RESOURCES
IN TWO AFRICAN CATCHMENTS

INTEGRATED MODELING
OF LAND AND WATER RESOURCES
IN TWO AFRICAN CATCHMENTS

DISSERTATION

Submitted in fulfillment of the requirements of
the Board for Doctorates of Delft University of Technology
and
of the Academic Board of the IHE-Delft
Institute for Water Education
for
the Degree of DOCTOR
to be defended in public on
Monday,19 March 2018, at 10:00 hours
in Delft, the Netherlands

by

Seleshi Getahun YALEW
MSc Water Science and Engineering, UNESCO-IHE, Delft, The Netherlands
BSc Information Systems, Addis Ababa University, Addis Ababa, Ethiopia
born in Dessie, Ethiopia

This dissertation has been approved by the
promotor: Prof.dr.ir. P. van der Zaag
promotor: Prof.dr.ir. A.B.K. van Griensven

Composition of the doctoral committee:

Chairman	Rector Magnificus TU Delft
Vice-Chairman	Rector IHE Delft
Prof.dr.ir. P. van der Zaag	IHE Delft / TU Delft, promotor
Prof.dr.ir. A.B.K. van Griensven	IHE Delft /Vrije Universiteit Brussels, promotor

Independent members:
Prof.dr.ir. N.C. van de Giesen	TU Delft
Prof.dr. W. Bewket Alemayehu	Addis Ababa University, Ethiopia
Prof.dr. A. Lehmann	University of Geneva, Switzerland
Dr. N. Deligiannis	Vrije Universiteit Brussels, Belgium
Prof.dr.ir. H.H.G. Savenije	TU Delft (reserve member)

CRC Press/Balkema is an imprint of the Taylor & Francis Group, an informa business

Published by:
CRC Press/Balkema
Schipholweg 107C, 2316 XC, Leiden, the Netherlands
Pub.NL@taylorandfrancis.com
www.crcpress.com – www.taylorandfrancis.com
ISBN 978-1-138-59338-1

To my mother

For her patience and understanding,

Because she always believes.

ማስታወሻቱ፤ ሊታጋሽና ጽኑ እናቴ ይሁንልኝ።

ACKNOWLEDGEMENTS

First and foremost, my sincere gratitude is due to Professor Pieter van der Zaag and Professor Ann van Griensven who have been there as my supervisors during this journey and have been unstinting in their support and constructive critique. Their choice of suggestion as opposed to dogma during their supervision has especially been my favorite and a constant inspiration for me to aspiring for independent research in integrated land and water resources research. Many thanks are also due to Dr. Marloes Mul and many other coauthors to my papers, some of which are part of this thesis, for positively contributing towards my growth in the realm of environmental research.

My heartfelt thanks and appreciation goes also to my wife, Lideta, and my children, Sophia and Alula, for different reasons: the earlier for giving me the space and the time to work on my dissertation, including the late nights and the weekends, and the latter for consistently reminding to creating laughter and fun.

Finally, I am grateful for the financial contribution by the AFRIMASON, AQUAREHAH, and NWO-WOTRO projects towards my PhD study. Similarly, I am thankful to all agencies and individuals who shared their local and regional environmental datasets.

SUMMARY

Land and water are two of the most important and interacting natural resources that are critical for human survival and development. Growing population and global economic expansion are accelerating the demand for land and water for uses such as agriculture, urbanization, irrigation, hydropower, and industrialization. The land surface changes dynamically due to these demands and other socio-economic drivers. Biophysical factors such as topographic suitability, climate change, and rainfall variability further influence land-use changes and land-use change decisions. Water resources are likewise experiencing pressure from overuse, pollution, and changes in hydrologic processes as a result of both socio-economic and biophysical factors.

Land and water resources are considered to have strong interactions. Although the science of this interaction is not new, the two are typically managed under separate governance systems (Le Maitre et al., 2014). As a result, land-use change and hydrology or water resources are studied usually separately. Modelers and model developers reproduced the existing separation in governance of land and water resources in their respective models, which gave rise to mild treatment of one of the resources when analyzing the other. The result is that hydrology and water resources are often considered to be processes and resources affected only by biophysical components, ignoring anthropogenic contributions that may actually influence such characteristics through direct effects on land-use, for instance.

The need to understand and explicitly represent the interaction between water resources on the one hand and land-use changes with its drivers on the other is imperative for sustainable management of integrated natural resources management in general and land and water resources management in particular. In recent years, interesting sub-disciplines such as 'socio-hydrology' are emerging which recognize and emphasize the importance of socio-economic and anthropogenic effects on hydrologic and water resources analysis. Interaction and feedback between land-use and water resources is still not explicitly and dynamically represented in most scientific modeling tools related to land and water, however. Furthermore, often due to limitations of access and computing resources, frameworks for communicating integrated assessment and modeling in a way that can be operational to resource managers and decision makers related to such resources is limited. This is especially evident in developing regions.

In this study, an integrated modeling approach is devised to modeling and testing interactions of land and water resources with focus areas in two selected basins in Africa: the Upper Blue Nile in Ethiopia, and the Thukela/Drakensberg in South Africa. The study focuses on analyzing land-use change drivers, assessing dynamic feedback between land-use and hydrology, and developing methods and tools for improved assessment and model representation in integrated modeling. For facilitating integrated modeling and analysis of land and water resources, various open-source tools and standards for simplified data access, computation and communication of results of integrated analysis are developed and tested.

First, land-use suitability was computed using spatial and biophysical indicators in the Blue Nile and the Thukela basins. Methods and techniques were developed for online assessment of land-suitability using global and open source datasets. Based on the land-use suitability indicators developed thus far and in light of various socio-environmental land-use change drivers, respective land-use change models were developed and parameterized for the two basins.

After the land-use models calibrated and evaluated, independent hydrologic models were developed and coupled with the land-use change models to test hydrologic response to varying land-use changes. In addition, methods to quantifying the effects of dynamic feedback between land-use change and hydrology on catchment ecosystem services were developed and tested.

Overall, results of this study showed that major changes in land-use have been observed in the past two to three decades in the study regions. The changes have generally resulted in expansion of agricultural lands at the expense of other land uses such as forest and grass lands. The main drivers and factors contributing to these changes in land use include increasing population and livestock, reduced distances from various infrastructures such as roads and markets/urban areas, and topographic factors such as slope, and land degradation. Model results of interactions of land-use change and hydrology/water resources showed that land-use change influences hydrologic response, demonstrated using stream flow responses. These influences are especially pronounced during high and low-flow seasons. Likewise, hydrologic processes and water resources availability are shown to influence land-use suitability and hence land-use change responses.

The study concludes that the investigation of socio-economic and biophysical land-use change drivers, and spatially explicit representation of feedback between land and water resources through respective models is imperative for informed policy and decision making in sustainable environmental management. However, such a holistic and integrated modelling, as much it is needed for sustainable land and water resources management, it is data demanding, complex and uncertainties can be high.

This study contributes to the field of integrated modeling for the sustainable management of natural resources in general and land and water resources management in particular, through:

1. Identifying and devising methods for analyzing land-use change drivers in catchments

2. Presenting methods for analyzing the dynamics of spatially explicit land-use suitability

3. Establishing the effects of dynamic feedback between land-use and water resource models for quantification of watershed ecosystem services.

4. Testing frameworks for simplified data access, computation and presentation of integrated environmental model results.

The study concludes with recommendations and suggestions for future research and improvement of limitations of the methods, approaches and tools used. Main areas for further research are:

1. Investigation of the feasibility of in-built and dynamic land-use change modules within hydrologic modeling frameworks or vice-versa to overcome complexity of coupling land-use change and hydrologic models.

2. Investigation of the applicability of global datasets and methods in supporting operational decisions though crop-specific land suitability assessment.

TABLE OF CONTENTS

LIST OF FIGURES

LIST OF TABLES

ACRONYMS

DEM	Digital Elevation Model
FAO	Food and Agriculture Organization
ha	Hectare
LULC	Land use/land cover
SRTM	Shuttle radar topographic Mission
SWAT	Soil and Water Assessment Tool
NSE	Nash-Sutcliffe efficiency
PBIAS	Percent bias
MCDA	Multi-criteria decision analysis
NMSA	National Meteorological Services Agency
MoWE	Ministry of Water and Energy of Ethiopia
LULC	Land use and land cover
mm/a	Millimeter per annum
m.a.s.l	Meters above sea level
ASCII	American Standard Code for Information Interchange
ES	Ecosystem services
INRM	Integrated natural resources management

Chapter 1. Introduction

Where water is boss, there the land must obey.
- African proverb

1.1 Background

Land-use and land cover (LULC) change has both resource management and strategic relevance (Aspinall and Justice, 2003). With respect to resource management, for instance, it impacts the sustainable use of water and other natural resources. On a strategic level, it can have profound implication on economic and development strategy of a community or a nation. LULC change may result in land degradation and soil erosion. Literatures have reported assessments of impacts of LULC on different hydrological events and components of catchments including discharge (Bewket and Sterk, 2005), flood runoff (Ashagrie et al., 2006; O'Connell et al., 2007; Ott and Uhlenbrook, 2004), groundwater (Harbor, 1994), and surface water quality (Tong and Chen, 2002). The studies demonstrated potentially significant effects of land-use change on various hydrological components of catchments and river basins. Land-use changes and decisions are also reported to be influenced by availability of water, among other things (Calder, 1998; Letcher et al., 2007; van Oel et al., 2010). However, research results on the effects of land-use and land- cover change on water resources or vice-versa vary greatly and the topic remains inconclusive (Zhou et al., 2015).

1.2 Integrated Assessment Modeling

Integrated assessment modeling is an approach that integrates knowledge from two or more scientific domains into a single framework for solving socio-environmental problems (Edmonds et al., 2012). The goal of integrated assessment modeling is to ensure that policy decisions are informed by a thorough understanding of the interdependencies and interactions within a system for sustainable socio-economic and environmental spheres. The growing interplay between social, economic, and environmental issues demands integrated policies (Easterling, 1997; Jakeman and Letcher, 2003; Rotmans and van Asselt, 1999). Integrated assessment of interactions between land and water resources use and management is critically important particularly in Africa due to multi-dimensional pressures on watersheds and river basins of the continent including population growth, climate variability, soil erosion, deforestation and land degradation. Current and projected demand for land and water resources has created upstream-downstream tensions on water uses on various parts of the continent. An integrated assessment of available and potential land and water resources, suitability of land for various uses and identification and projections of hotspot locations for various ecosystem services is an important endeavor towards a regulated use of these resources. Even though integrated assessment modeling is an essential way of doing socio-environmental science (Harris, 2002; Jakeman and Letcher, 2003), the number of scientific domains involved can make communication of results of such models rather difficult (Liu et al., 2008). A simplified platform that can translate complex assessment results for regional and local policy makers should be taken as an important part of the endeavor towards making integrated assessment modeling more impacting.

1.3   Study Areas

Two study locations, the Thukela catchment in the Drakensberg, South Africa, and the Abbay/Upper Blue Nile basin in Ethiopia, were selected for this study. The selected locations are part of the EU/FP7 funded AFROMAISON[1] project whose objective was to use integrated assessment modeling methods and tools to identifying, valuing, mapping and managing ecosystem services in relation to integrated land and water resources in five meso-scale catchments in Africa.

The Abbay basin

Located in the western part of Ethiopia (Figure 1-1), Abbay is the most important river basin in Ethiopia by most criteria: it accounts for about 20% of the nation's land area; 50% of its total average annual runoff; 25% of its population; and over 40% of its agricultural production (EEPCo, 2014). The basin covers an area nearly 200,000 km^2 and is important not only in Ethiopia, but also for its significant contribution of runoff and fertile soil to Sudan and Egypt downstream. Anthropogenic factors combined with torrential runoffs in the rugged highlands in the basin have caused considerable land degradation and soil erosion in upstream catchments, whereas deforestation is a major concern in the mid and low lands of the basin (Gebrehiwot et al., 2014; Hurni et al., 2005).

---

[1] http://www.afromaison.net/

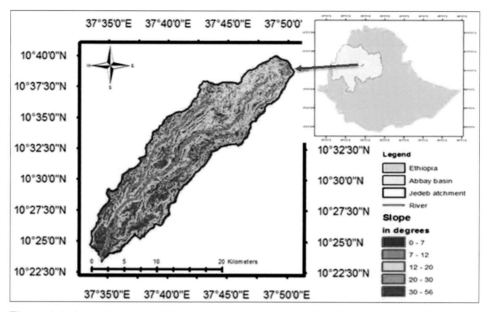

Figure 1-1. Location map of the Jedeb catchment and the Abbay basin in Ethiopia

Research shows that besides diverse socio-economic and biophysical composition over an extended area, intensified land use and increasing population led to increased land degradation and soil erosion from the Ethiopian highlands leading to high sediment loads in the Upper Blue Nile River Basin (Bewket and Sterk, 2002; Hurni et al., 2005). Poor land-use practices and management in the region are pointed as some of the main causes of high soil erosion rates and loss of agricultural nutrients in the basin (Setegn et al., 2008).

The basin has a heterogeneous climate ranging from humid to semi-arid which is mainly dominated by its near equatorial location (between latitudes of 70 45'N and 120 45'N) and differences in elevation (altitudes of 500 m to more than 4,200 m.a.s.l.). There are three seasons defined by the National Meteorological Services Agency (NMSA) of Ethiopia: Kiremt, Bega and Belg. Kiremt is the rainy season from June to September, and Bega is the dry season from October to January, whereas Belg is the "short rains" season from February to May. The river system drains a highly seasonal residual rainfall off the Ethiopian highlands mainly during the months June to September (the Kiremt Season). According to data from the Ministry of Water and Energy of Ethiopia (MoWE), a minimum and maximum temperature of 11.4 and 25.5°C, respectively; a minimum, maximum and average annual rainfall of 800, 2220 and 1420 mm/a respectively; and an average annual potential evaporation of 1300 mm/a characterize the basin.

The Jedeb catchment in the Abbay basin is used as a focus case study catchment in this study. The catchment is situated on the upstream highland of the Abbay basin where the land has been used for farming and grazing for hundreds of years. With area coverage of 297km^2, the catchment is dominated by agriculture. No prevalent forest or natural vegetation exists except for shrub lands and alpine grasslands in the mountains. In addition to torrential runoff washing off the rugged terrains, poor land-use management has been reported as a cause for gully formation, soil erosion and land degradation that threatens the livelihood of subsistence farmers (Kotch et al., 2012; Smit et al., 2017; Tekleab et al., 2014a; Yalew et al., 2012). In addition to poor land, soil and water resources management, intensive and extensive agricultural practice to meet the growing population need in the highlands is becoming unsustainable (Bewket and Teferi, 2009; Desta, 2000).

The Thukela catchment

The district of Thukela is located in the upper part of the Thukela basin located within the province of KwaZulu-Natal, South Africa (Figure 1-2). This district is predominantly rural, characterized by socio-economic indicators such as low revenue base, poor infrastructure, limited access to services, and low economic base. Among the most substantial pressures to ecosystems in the study area are intensive livestock grazing and poor land management practices, leading to soil loss and land degradation. The district had a population of about 670,000 inhabitants in 2012, resulting in a population density of 60 people per km^2, with a slightly increasing trend. The grassland biome forms a large and important component of South African vegetation (Scott-Shaw and Schulze, 2013). Livestock holds major economic and social values for the communal farmers in the country. Beyond its economic benefit, livestock is used as a sign of prestige in the community. Grazing land, thus, is of particular importance to the rural community at large.

---

[2] http://www.statssa.gov.za/Census2011/Products.asp

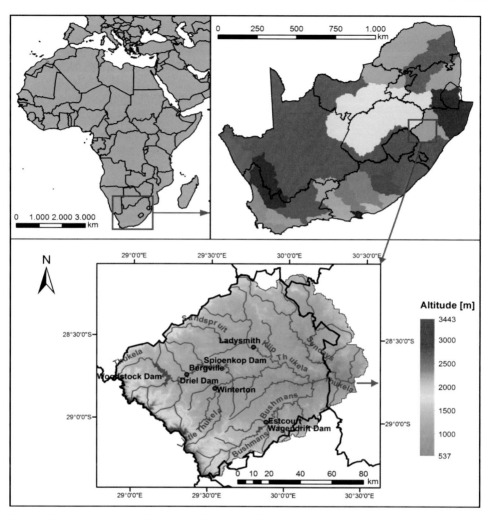

Figure 1-2. Location and topographic map of the study area

Black lines in the top right map indicate borders of the provinces of South Africa, Lesotho and Swaziland; different colors distinguish primary catchments, i.e. the primary drainage regions. The lower map defines the study area, layers comprising Thukela displays altitudes, including major rivers, dams, and cities, whereas thick lines represent borders of the district municipalities. The blue arrow indicates the catchment outlet at the town of Tugela Ferry.

1.4   Problem Descriptions

Problems in the Abbay basin

Whereas deforestation is indicated as a major concern in the mid and low lands of the Abbay basin, anthropogenic factors combined with biophysical factors such as torrential runoffs in the rugged highlands of the basin have caused considerable land degradation and soil erosion in the upstream (Gebrehiwot et al., 2014; Hurni et al., 2005). In addition to poor land, soil and water resources management, intensive/extensive agricultural practice to meet the needs of growing population especially on highlands of the basin is becoming unsustainable (Bewket and Teferi, 2009; Desta, 2000). Furthermore, the intensive land uses in the upstream have been cited as a cause for sedimentation and reservoir storage problems on downstream dams (Awulachew et al., 2009; Betrie et al., 2011; Hurni et al., 2005). Continuous overuse of the agricultural land in the densely populated highlands has resulted in high gully formations, soil erosion and land degradation problems in the Abbay basin in general and in the Jedeb catchment in particular (Smit et al., 2017). Despite the fact that the Abbay basin is an important area of research interest due mainly to its heterogeneous hydro-climatology, socio-economic as well as biophysical diversity, efforts for assessing drivers of changes and integrated assessment modeling of socio-economic and biophysical dynamics with regards to implications on land and water resources management in the basin is still limited.

Controversies on whether water resource development projects in the upstream of the Nile basin will generate positive externalities both within or outside the basin call for a framework that enables an integrated analysis of land and water dynamics in the region (Goor et al., 2010). As such, close investigation of trends of land and water resources in the basin, analysis of what drives land-use changes especially in the upstream, and parameterization of models that can spatially explicitly project trajectory of changes in the basin can be valuable inputs to local and regional policy and decision makers.

Problems in the Thukela catchment

South Africa's grassland biome has been identified as critically endangered (Neke and Du Plessis, 2004; Olson and Dinerstein, 1998; Reyers et al., 2001). High livestock population and poor soil and land management causes degradation of the grasslands. Increasing demand for arable and urban land decreases the extent of the grazing lands. This is further influenced by changes in climate, and has multiple impacts, such as increased erosion and changes in flow regime. Overgrazing, agricultural and urban expansion, mining, and poor land-use management practices have been reported to have resulted in land and soil degradation in the Thukela catchment. It has been noted that poorly managed livestock grazing can lead to the emergence of three regional syndromes inherent to global grazing: desertification, woody encroachment and deforestation (Asner et al., 2004; Mabbutt, 1984). Identification, assessment and mapping of grassland biomass that can be used for sustainable grazing by livestock in the study area are essential for long term rangeland policy and agro-hydrological decision making in the catchment.

1.5   Research Objectives

The general objective of this study is to develop and test novel approaches and methods for analyzing feedbacks between land-use change and hydrologic models for integrated and sustainable watershed management. Specific objectives are:

1. Investigating the dominant biophysical and human drivers of land-use change and their implications on water resources

    o   Parameterization of a land-use change model based on analysis of land-use trends, practices, and land use and land cover (LULC) change drivers in the Abbay basin.

    o   Spatiotemporal analysis of dynamic land suitability in the basin

2. Developing and testing methods to assess hydrologic impact of land-use change

    o   Investigating impacts of 'semi-dynamic' land-use inputs on streamflow response in the Abbay basin

3. Developing and testing methods for quantifying effects of the interaction between land-use change and hydrology on catchment ecosystem services.

    o   Coupling of land-use and hydrologic models to analyze impacts of the dynamic interaction between land-use and hydrology.

    o   Testing effects of dynamic feedback between coupled land-use change and hydrologic models on quantification of catchment ecosystem services.

4. Developing and testing a framework/prototype for easy access, computation and communication of integrated assessments for regional environmental decisions.

   o Development of a web-based framework for accessing, computation and visualization of regional land-use suitability based on globally accessible open source platforms.

1.6 Methodology

Often times, socio-environmental problems cross boundaries between academic disciplines. Integrated assessment modeling is an increasingly common approach to socio-environmental problem assessment that involves knowledge from two or more scientific disciplines (Jakeman and Letcher, 2003; Rotmans and van Asselt, 1999). This study adopts integrated assessment modeling approaches and methods with emphasis on modeling the effects of interactions of land-use change and hydrologic feedback. In line with integrated assessment modeling, the study develops tools and web-based frameworks for accessing, computing and communicating integrated land and water resources analysis based on global data sources and open-source computing platforms.

1.7 Structure of the thesis

The thesis is organized in two parts and 7 chapters. A general introduction and overview of the motivations, locations, problem descriptions, objectives and methodology of the overall study are presented in this introductory chapter. Chapters 2, 3 and 4 (Part I) investigate land-use suitability assessment methods and tools and land-use change modeling in the context of the Abbay basin. Chapters 5 and 6 (Part II) investigate modeling of dynamic and 'semi-dynamic' feedback between land-use change and hydrologic models with case studies in the Abbay (Ethiopia) and the Thukela (South Africa) basins. Chapter 7 presents the conclusion, limitation and recommendation of the overall study. Appendices present supplementary materials.

PART I: LAND-USE CHANGE MODELING

We live in the present, we dream of the future, but we learn eternal truths from the past.

- Madame Chiang (b.1898), educator, reformer

Chapter 2. Land suitability assessment in the Abbay basin[3]

2.1 Introduction

Increasing population and the associated growing demand for food and other agricultural commodities have caused an intensification and extensification of the agricultural sector witnessed in the last decade (Lambin and Meyfroidt, 2011; Rudel et al., 2009; Tscharntke et al., 2012). As an agriculture dominated basin in Ethiopia, the Upper Blue Nile seems to be experiencing similar pleasures (Bewket and Sterk, 2002; Gebrehiwot et al., 2014). However, the amount, location and degree of suitability of the basin for agriculture do not seem well studied and/or documented (Yalew et al., 2016c). Haphazard land-use has thus far resulted in continuing deforestation, exhaustion of soil fertility, increased soil erosion and land degradation especially in the basin's highland catchments (Awulachew et al., 2010; Bewket, 2002; Zeleke and Hurni, 2001). Land suitability analysis can help establish strategies to increase agricultural productivity (Pramanik, 2016) by identifying inherent and potential capabilities of land for intended objectives (Bandyopadhyay et al., 2009). It can also help identify priority areas for potential management and/or policy interventions through land and/or soil restoration programs, for instance.

The Analytical Hierarchy Process (AHP) (Saaty, 1980) technique integrated with GIS application environments has been used for agricultural land suitability analysis on various case study sites around the world (Akıncı et al., 2013; Malczewski, 2004; Pramanik, 2016; Zabihi et al., 2015; Zolekar and Bhagat, 2015). It involves pair-wise and weighted multi-criteria analysis on a number of selected socio-economic and biophysical drivers. The technique has extensively been used for land suitability analysis at local and region levels for watershed planning (Steiner et al., 2000), vegetation (Zolekar and Bhagat, 2015) and agriculture (Akıncı et al., 2013; Bandyopadhyay et al., 2009; Motuma et al., 2016; Shalaby et al., 2006). Biophysical parameters such as land cover, slope, elevation, and soil properties such as depth, moisture, texture and group are frequently used for assessment of land suitability evaluation (Brinkman and Young, 1976; Zolekar and Bhagat, 2015). 'Expert

---

[3] This chapter is based on a paper published on the journal of Modeling Earth Systems and Environment:
Seleshi G. Yalew, Ann van Griensven, Marloes L. Mul, Pieter van der Zaag (2016). Land suitability analysis for agriculture in the Abbay basin using remote sensing, GIS and AHP techniques. Model. Earth Syst. Environ. (2016) 2: 101.

opinion' is used for weighting such factors in influencing land suitability through pair-wise comparison in AHP.

In this chapter, we analyzed agricultural land-use suitability in the Abbay basin using AHP and GIS based weighted overlay analysis (WOA) techniques. Multiple criteria for agricultural land-use suitability mapping were derived based on literature reviews, field investigations and following FAO guidelines for agricultural land-use evaluation (Bandyopadhyay et al., 2009; Brinkman and Young, 1976; Zabihi et al., 2015; Zolekar and Bhagat, 2015). Identification and mapping of agricultural land suitability is especially important in the basin given the following considerations: (i) the pressing need to increase agricultural productivity to meet growing food demands, (ii) the growing risks of increased rainfall variability due to climate change in already water limited agricultural systems, and (iii) the growing interest by local and regional policy and management bodies for evaluation of land capability for various land-use alternatives.

2.2   Study Area

This analysis was carried out on the Abbay basin and then downscaled to the Jedeb catchment in the basin for visual inspection and evaluation of the use of global datasets on local analysis of land-use suitability. Detailed description of the basin is presented in chapter 1, and sub-basins are shown in Figure 2-1.

Figure 2-1. The Abby basin and its sub-basins

2.3   Materials and Methods

Using literatures and guidelines on land evaluation for agriculture (Bandyopadhyay et al., 2009; Bojórquez-Tapia et al., 2001; Brinkman and Young, 1976; Olaniyi et al., 2015; Prakash, 2003; Wang, 1994; Zhang et al., 2015) we identified 9 important criteria that determine agricultural land suitability in the basin: soil type, soil depth, soil water content, soil stoniness, slope, elevation and proximity to towns, roads and water sources (Table 2-1). GIS raster datasets on each of these indicators were gathered and processed from several sources for the study areas. According to FAO guidelines (Brinkman and Young, 1976), land suitability for agriculture can be classified into 5 categories: (i) highly suitable (ii) moderately suitable (iii) marginally suitable (iv) currently unsuitable and (v) permanently unsuitable. In this study, we customized and reclassified each raster criteria layer into 4 categories with associated suitability score of 1 to 4 (4= highly suitable; 3=moderately suitable; 2=marginally suitable; and 1= unsuitable). The 'unsuitable' category represents the 'permanently unsuitable' category of FAO. Similar to what is defined as 'currently unsuitable' in the FAO method, we excluded forest and protected areas from the suitability computation altogether assuming that such land may not be used (and hence 'unavailable') for agriculture in favor of other ecological services (biodiversity conservation). Weights for each of the selected criterion were calculated using the AHP technique. After the weight of each raster dataset was computed, a GIS based WOA was carried out to establish a suitability map of the basin. The process diagram of the method used for suitability analysis in this chapter is shown in Figure 2-2.

Table 2-1. Data and data sources

Data	Spatial resolution	Source
Elevation	30m	SRTM [4]
Slope	30m	Computed from SRTM
Soil type	5 arc minute	FAO (FGGD) (2013)
Soil depth	5 arc minute	FAO (2014)
Soil stoniness	1km	ISRIC-worldgrid1km (2014)
Soil water content	30 arc seconds	CGIAR-CSI ((2010)
Towns	Woreda (county) level	CSA (2007), FAO
Roads	All weather roads	CSA (2007), FAO
River/water bodies	Perennial streams	MoWR, Ethiopia
Protected areas	2.5 arc minute	IUCN & UNEP-WCMC (2012)
Land cover	300m	ENVISAT/MERIS (Bontemps et al., 2011)

---

[4] *Google Earth Engine*

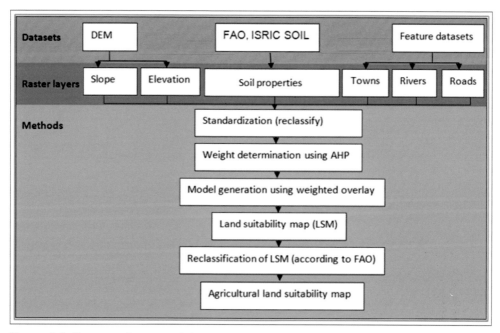

Figure 2-2. Process diagram of the methods

Generation of criteria maps

2.3.1.1 Slope and elevation

Slope and elevation data layers (Figure 2-3(a) and (b), respectively) were generated from the Shuttle Radar Topography Mission (SRTM) digital elevation model (DEM) data of 30 meter resolution available on Google Earth Engine (Gorelick, 2013; Gorelick et al., 2017). Based on FAO manual for agricultural watershed management (Sheng, 1990), agricultural suitability of different slope classes for the study area are defined as in Table 2-2. However, since no specific crop suitability is assumed, elevation value lower than 3,700 m a.s.l. is taken to be suitable for agriculture. Elevation above 3,700m is classified as 'high wurch' (frosty-alpine) and thus unsuitable for agricultural purposes according to the agro-ecological zoning of Ethiopia (FAO, 2003).

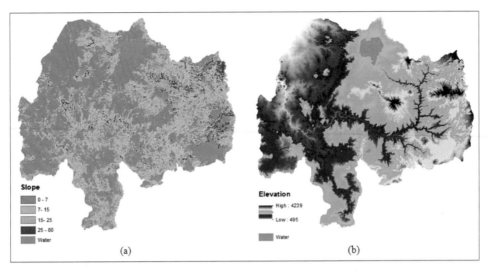

Figure 2-3. (a) Slope and (b) elevation maps of the Abbay basin

Table 2-2. Slope classes for agricultural suitability

Slope class (degree)	Suitability score
0-7	4
7-15	3
15-25	2
>25	1

2.3.1.2 Soil properties

Soil characteristics are one of the most important factors in agricultural land-use assessment (Bonfante and Bouma, 2015; Dominati et al., 2016; Juhos et al., 2016). In this study, soil depth, soil water content, soil type and soil stoniness are taken as indicators to assess general soil suitability for agriculture. Soil depth and averaged soil water content maps are shown in Figure 2-4. Soil type and soil stoniness are shown in Figure 2-5. The soil properties used here were standardized for land suitability assessment as shown in Table 2-3. These soil characteristics were categorized based on the soil classification and characterization guide for agricultural suitability by FAO (Sheng, 1990), and other guidelines for common biophysical criteria for defining natural constraints for agriculture (Van Orshoven et al., 2012).

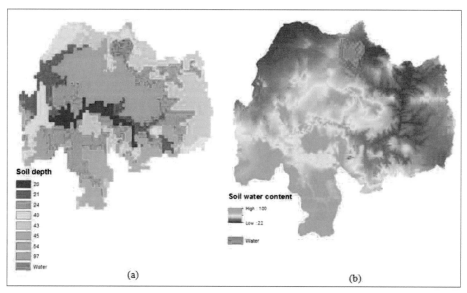

Figure 2-4. (a) Soil depth and (b) soil water content in the Abbay basin

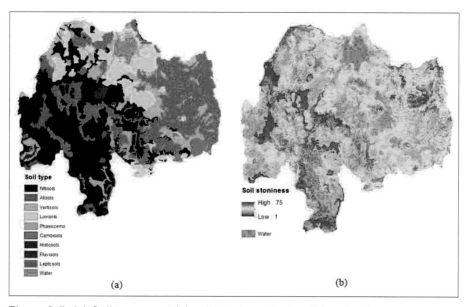

Figure 2-5. (a) Soil types and (b) soil stoniness in the Abbay basin

Soil stoniness refers to percentage of gravel/stone content within the top 90cm soil depth. Soil groups were classified based on their suitability and limitations for agriculture as outlined by FAO and the international livestock research institute (ILRI) (1992). Based on the guides, the major soils in the study area are classified as:

- Soils with very high potential: Nitisols (NS), Luvisols (LS), Cambisols (CS), Phaeozems (PS)

- Soils with few limitations for agriculture: Vertisols (VS), Alisols (AS)

- Soils with major limitations (low production potential, rocky terrain soils, poorly drained soils): Histosols (HS), Liptosols (LpS)

Soil water content dataset was derived from the spatially distributed soil-water balance model by Trabucco and Zomer (2010). In their model, Trabucco and Zomer simulated a soil-water balance model for the years from 1950-2000 as a height of water (in mm) per month (m) using Eq. 2.1.

$$\Delta SWC_m = EPr_{ec} - \Delta AET_m - R_m \qquad (2\text{-}1)$$

Where, ΔSWC_m is the change in soil water content, EPr_{ec} is the effective precipitation, ΔAET_m is the actual evapotranspiration, and R_m is the runoff component which includes both surface runoff and subsurface drainage. Furthermore, SWC_m may not exceed SWC_{max}, which is the total maximum soil water content (SWC) available in the soil for evapotranspiration. SWC_{max} was assumed by the modelers at a fixed spatial value of 350mm, which corresponds to average soil texture for a plant rooting depth of 2 meters. The soil water content was then computed as a linear percentage function of   actual and potential (maximum) soil water content over the months and the years from 1950-2000 as shown with Eq. 2.2.

$$\text{Ksoil}_m = \sum_{k=1}^{y} \left(\sum_{k=1}^{12} (SWC_m / SWC_{\max}) \right) \qquad (2\text{-}2)$$

Where, Ksoil_m is percentage of average soil water content, SWC_m is actual soil water content in month (m), $SWC_{\max}$) is the maximum (potential) soil water content, and y is year.

Table 2-3. Soil characteristics and suitability for agriculture

Soil property	Suitability score			
	4	3	2	1
Soil depth (cm)	<90	50-90	20-50	0-20
Soil Stoniness (%)	0-3	3-15	15-50	>50
Soil type	NS,LS,CS,PS	VS,AS	HS, LpS	---
Soil water content (%)	90-100	70-90	30-70	<30

2.3.1.3 Proximity to water, road and towns

Spatial proximities to water sources, road and towns (Figure 2-6 (a), (b), and (c), respectively) were computed using spatial overlay of respective GIS layers. Influences of distance parameters on agricultural land suitability, Table 2-4, were estimated based on literature and field observation (Bizuwerk et al., 2005; Wale et al., 2013).

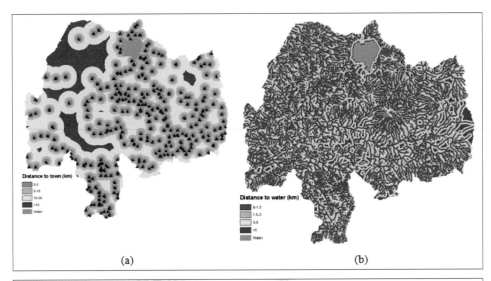

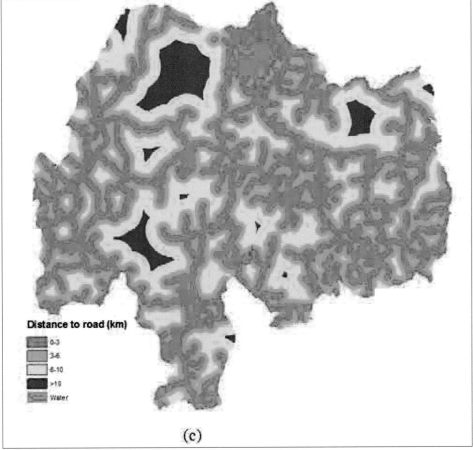

Figure 2-6. Distances to (a) town (b) water sources and (c) road in the Abbay basin

Table 2-4. Proximity influences on agricultural land suitability

Proximity (km)	Suitability score			
	4	**3**	**2**	**1**
Distance to town	0-5	5-10	10-30	>30
Distance to roads	0-3	3-6	6-10	>10
Distance to water	0-1.5	1.5-3	3-5	>5

In addition to the criteria inputs for agricultural land suitability assessment discussed thus far, data on land cover and protected sites where collected for overlay analysis to serve as constraint layers on the final suitability map. Land cover map (Figure 2-7a) was derived and reclassified from GlobCover2009. GlobCover2009 is a global land cover map based on ENVISAT's Medium Resolution Imaging Spectrometer (MERIS) Level 1B data acquired in full resolution mode with a spatial resolution of 300 meters (Bontemps et al., 2011). Map on protected sites (Figure 2-7b) which includes areas such as national parks and reserve sites was derived from the 'Protected Sites' global dataset of UNEP's World Conservation Monitoring Centre (UNEP-WCMC, 2012).

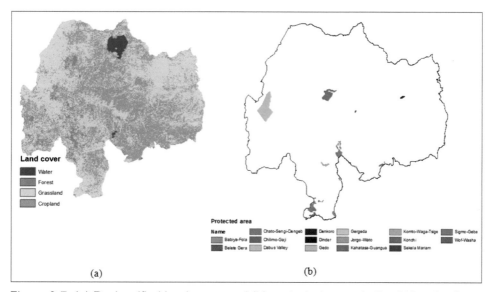

(a)   (b)

Figure 2-7. (a) Reclassified land cover and (b) protected areas in the Abbay basin

We assumed that forest, protected areas and water bodies as unavailable (and hence 'currently unsuitable') for agriculture. Changes in policy or management could easily change the suitability of these layers. A forest may, for instance, be deforested for large scale agriculture and thus changing its land suitability. These layers (Figure 2-8) are therefore used as constraints that are superimposed on top of the computed suitability map.

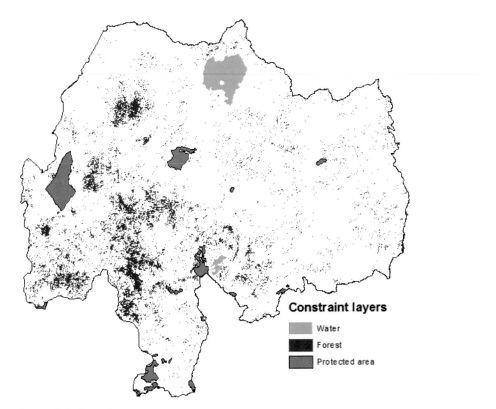

Figure 2-8. Constraint layers

Standardization of criteria maps

The selected criteria maps are initially in different units. For executing WOA for land suitability, the criteria maps need to be converted into a similar scale through standardization techniques. Standardization techniques convert the measurements in each criteria map into uniform measurement scale so that the resulting maps lose their dimension along with their measurement unit (Reshmidevi et al., 2009; Zabihi et al., 2015). For standardization, all the criteria vector maps were converted to raster data formats. The raster maps were then reclassified using the Spatial Analyst tool in ArcMap into 4 comparative categories as discussed earlier: Highly Suitable, Moderately Suitable, Marginally Suitable and Unsuitable. Once all the criteria maps are standardized, weights of each criteria map can be calculated using AHP. Then WOA method will be applied to produce the final suitability map.

Calculation of weights for criteria maps

The analytic hierarchy process (AHP) is used to calculate weights for the criteria maps. It is a structured method for analyzing complex decisions by breaking them into pair-wise alternatives of two at a time (Saaty, 1988, 2008). It involves sub-dividing big and intangible decision problems into minute sub-problems amenable for pair-wise comparison (Saaty, 1987). An AHP plugin tool for the ArcGIS environment (Marinoni, 2004; 2009) was used to compute weights for the different criteria layers. Using the pair-wise comparison matrix, the analytic hierarchy process calculates comparative weights for individual criterion layers. It also produces consistency ratio (CR) that serves as a measure of logical inconsistency of expert/user judgments during pair-wise criteria comparisons, measured using Eq. 2.3.

$$CR = \frac{CI}{RI} \qquad\qquad (2\text{-}3)$$

Where, CI represents consistency index, and RR represents random index.

The CR measurement facilitates identification of potential errors and thus judgment improvements depend on these values. According to Saaty (1988), if the CR value is much in excess of 0.1, the judgments during pair-wise comparison are untrustworthy because they are too close for randomness. Saaty (1988) provided a 'fundamental scale' for computing pair-wise comparison matrix of the criteria layers while performing an AHP (Table 2-5). This involves a construction of a matrix where each criterion is compared with the other criteria, relative to its importance, on a scale from 1 to 9. Scale 1 indicates equal preference between a pair of criteria layers whereas 9 indicates a particular criteria layer is extremely favored over the other during expert judgment (Malczewski, 2004; Saaty, 1988).

Table 2-5. The fundamental scale for pair-wise comparison matrix (Saaty 1980)

Relative Importance	Definition	Description
1	Equally important	Two criteria enrich equally to the objective
3	Slightly important	Judgments and experience slightly favour one criteria over another
5	Fundamentally important	Judgments and experience strongly favour one over the other
7	Really important	One is strongly favoured and its dominance established in practice
9	Absolutely important	Evidence favouring one criteria over another is of the highest probable order of affirmation
2,4,6,8	Adjacent	Used when intermediate importance is needed

Reciprocals: If criteria i has one of the above numbers designated to it when compared with criteria j, then j has the reciprocal value when compared with i (see Table 2-6).

After determining the relative importance of each criteria layer, through pair-wise comparison matrix, these values are entered on an ArcGIS based AHP tool to produce associated weights and CR value. Table 2-6 shows inputs to the pair-wise comparison for the AHP analysis to compute weights for the criteria layers. The weights produced from the AHP procedure using inputs in this table range between 0 and 1, where 0 denotes the least important and 1 the most important criteria determination of land suitability. The consistency of the pair-wise comparisons for the computation of criteria weights is shown in Table 2-7.

Table 2-6. Inputs to the AHP for the pair-wise comparison analysis and computation of weights

Criteria layer	Soil water content	Soil stoniness	Soil type	Dist. to water	Elevation	Slope	Soil depth	Dist. to road	Dist. to town	Criteria Weight
Soil water content	1	3	7	8	9	2	4	5	6	0.307
Soil stoniness	0.33	1	5	6	7	1	2	3	4	0.154
Soil type	0.14	0.2	1	2	3	0.17	0.25	0.33	0.5	0.037
Dist. to water	0.13	0.17	0.5	1	2	0.14	0.2	0.25	0.33	0.026
Elevation	0.11	0.14	0.33	0.5	1	0.13	0.17	0.2	0.25	0.019
Slope	0.5	2	6	7	8	1	3	4	5	0.218
Soil depth	0.25	0.5	4	5	6	0.33	1	2	3	0.109
Dist. to road	0.2	0.33	3	4	5	0.25	0.5	1	2	0.076
Dist. to town	0.17	0.25	2	3	4	0.2	0.33	0.5	1	0.053

Table 2-7. Indices computed using the GIS based AHP tool

Index	Value
Consistency Index (CI)	0.0495
Random Index (RI)	1.46
Consistency Ratio (CR)	0.0339

WOA

After computation of weights for each raster layer using AHP, weighted overlay analysis (WOA) is performed on an ArcGIS environment. Weighted overlay is an intersection of standardized and differently weighted layers during suitability analysis (Zolekar and Bhagat, 2015). The weights quantify the relative importance of the suitability criteria considered. The suitability scores assigned for the sub-criteria within each criteria layer were multiplied with the weights assigned for each criterion to calculate the final suitability map using the WOA technique (see Eq. 2.4).

$$S = \sum_{i=1}^{n} W_i X_i \qquad\qquad (2\text{-}4)$$

where S is the total suitability score, W_i is the weight of the selected suitability criteria layer, X_i is the assigned sub-criteria score of suitability criteria layer i, and n is the total number of suitability criteria layer (Cengiz and Akbulak, 2009; Pramanik, 2016).

2.4   Results

The weighted overlay analysis carried out using the criteria layers with their respective weights generated a combined suitability map (Figure 2-9). Forest, protected area and water bodies were computed and then superimposed on this suitability map to determine the final suitability map (Figure 2-10). According to this map, it was determined that 28.6% (57,050 km^2) of the study area is highly suitable for agriculture, 48.9% (97,812 km^2) is moderately suitable, and 6.2 % (12,378 km^2) is marginally suitable for agriculture. About 6% (11,978 km^2) is determined to be 'unsuitable' whereas the rest 10.3% (20,594 km^2) is determined unavailable (or currently unsuitable) categories (see Table 2-8).

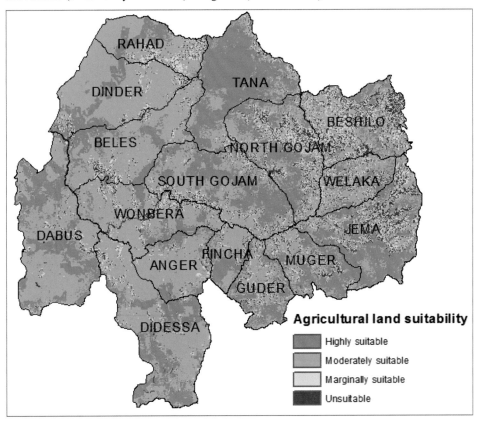

Figure 2-9. Agricultural land suitability in the Abbay basin excluding constraint layers

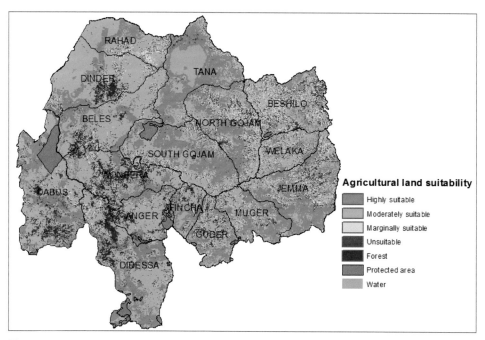

Figure 2-10. Agricultural land suitability in the Abbay basin with constraint layers.

Water bodies, forest cover and protected areas are treated as 'unavailable' or are constraints for the suitability analysis and are instead superimposed on the final suitability map (Figure 2-10).

Table 2-8. Summary of agricultural land suitability map of the Abbay basin

Suitability	Area (km^2)	Percent (% of the basin area)
Highly suitable	57,050	28.6
Moderately suitable	97,812	48.9
Marginally suitable	12,378	6.2
Unsuitable	11,978	6
Unavailable*	20,594	10.3
Total area	**199,812**	**100%**

* This includes protected areas, forest cover and water bodies.

The land-use suitability of the basin was further quantified per catchments in the basin (Table 2-9.

Table 2-9. Summary of land suitability for agriculture in catchments of the Abbay basin [km2 (%)]*

No	Catch.	Total area	Highly suitable	Moderately suitable	Marginally suitable	Unsuitable	Unavailable
1	Tana	15054	7535 (50.1)	3651 (24.3)	246 (1.6)	378 (2.5)	3244 (21.5)
2	N. Gojam	14389	4539 (31.5)	6164 (42.8)	1750 (12.2)	1607 (11.2)	329 (2.3)
3	Beshilo	13242	1582 (11.9)	5959 (45)	2526 (19.1)	2954 (22.3)	221 (1.7)
4	Welaka	6415	1061 (16.6)	3959 (61.7)	722 (11.3)	550 (8.6)	113 (1.8)
5	Jema	15782	6301 (39.9)	5524 (35)	1750 (11.1)	1961 (12.4)	246 (1.6)
6	S.Gojam	16762	6516 (38.9)	7514 (44.8)	869 (5.2)	777 (4.6)	1086 (6.5)
7	Muger	8188	3573 (43.6)	3470 (42.4)	511 (6.2)	457 (5.6)	177 (2.2)
8	Guder	7011	2299 (32.8)	3224 (46)	499 (7.1)	383 (5.5)	606 (8.6)
9	Fincha	4089	899 (22)	2113 (51.7)	108 (2.6)	192 (4.7)	777 (19)
10	Didessa	19630	6129 (31.2)	9761 (49.7)	467 (2.4)	300 (1.5)	2973 (15.2)
11	Anger	7901	1283 (16.2)	4819 (61)	123 (1.6)	128 (1.6)	1548 (19.6)
12	Wombera	12957	1892 (14.6)	6916 (53.4)	654 (5)	713 (5.5)	2782 (21.5)
13	Dabus	21032	6458 (30.7)	9762 (46.4)	315 (1.5)	315 (1.5)	4182 (19.9)
14	Beles	14200	2978 (21)	9251 (65.1)	531 (3.7)	423 (3)	1017 (7.2)
15	Dinder	14891	2344 (15.7)	10419 (70)	521 (3.5)	383 (2.6)	1224 (8.2)
16	Rahad	8269	1651 (20)	5306 (64.2)	786 (9.5)	457 (5.5)	69 (0.8)
	**Max:		50.1	70.0	19.1	22.3	21.5
	Min:		11.9	24.3	1.5	1.5	0.8
	Mean:		27.3	50.2	6.5	6.2	9.9
	StDev:		11.7	12.0	5.0	5.4	8.2

*Area of the catchment in km² and in bracket percentage area of the catchment.

** Max, Min, Mean and StDev denote maximum, minimum, mean and standard deviation statistics, respectively, of percentage of area coverage in the 16 catchments.

For a closer evaluation of the suitability analysis at a smaller scale, and as a basis for the land-use change modeling in the coming chapters, we zoomed into the Jedeb catchment in the basin and investigated the agricultural suitability out of the basin scale suitability map for 2009, Figure 2-11. Previous land cover studies of this catchment based on Landsat imagery show that agricultural land (cropland and plantation) covered 69.5% of the catchment in 2009 (Teferi et al., 2013b; Yalew et al., 2012). Analysis of the reclassified MODIS land cover map in this study shows a closer areal coverage of agricultural land at 71% for the same year.

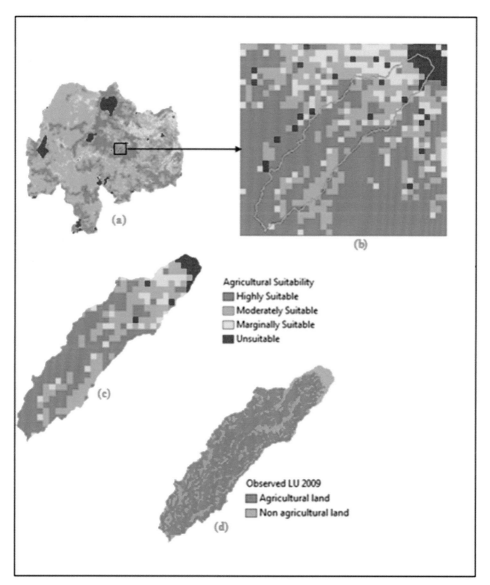

Figure 2-11. The Jedeb catchment in the Abbay basin (a,b); suitable agricultural land of 2009 (c), and reference agricultural land cover map of 2009 (d).

Taking the Landsat based classified agricultural land cover for 2009 as a reference dataset, spatial difference was computed between the two maps to assess where, and whether, suitability categories may match the observed agricultural land in the catchment, Figure 2-12.

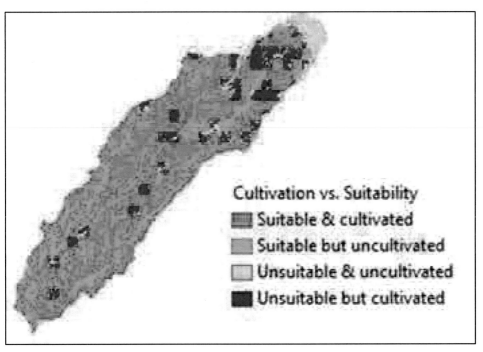

Figure 2-12. Spatial difference between the reference and the suitability maps for agricultural land-use of 2009

Table 2-10. Observed agricultural land-use vs. suitable agricultural land in the Jedeb catchment for 2009

Cultivation vs. Suitability	Amount (%)
Suitable & cultivated	61.3
Suitable but uncultivated	22.2
Unsuitable & uncultivated	7.6
Unsuitable but cultivated	8.9

2.5   Discussion

A closer look at the percentage coverage of the suitable lands in the catchments shows high variation between the different catchments of the basin (see Table 2-9). About 50% of the Tana catchment (North), for instance, is classified as 'highly suitable' whereas as low as only 12% is classified in the same category in Beshilo (North-East). Similarly, there is a large variation in percentage coverage of 'moderately suitable' lands per catchment which ranges from 70% in the Dinder catchment (North-West) to about 24% in the Tana catchment. A much lower variation in percentage area between catchments is seen when considering the sum of 'highly suitable' and 'moderately suitable' categories (Max=86%; Min=57%; Mean =77%; Stdev=7) compared with the sum of 'marginally suitable' and 'unsuitable' lands (Max=41%; Min=3%; Mean=12; Stdev=10).

What is generally noticeable is that the North, North-West, South and South-West catchments of the basin seem to have larger percentage area for 'highly suitable' and 'moderately suitable' land for agriculture. On the other hand, the Western, North-Western and Central highlands of the basin seem to have higher coverage of 'marginally suitable and 'unsuitable' lands for agriculture. Looking at some of the main factors for weight computation in the AHP analysis such as slope, soil water content and soil stoniness, it is easy to see that the North-Western and central highlands are dominated by steep slope ranges (25-80 degrees, Figure 2-3a), low percentage of soil water content (22%, Figure 2-4b) and high level of soil stoniness (75%, Figure 2-5b). This part of the basin is also located on a relatively higher elevation range (3000-4239 m.a.s.l.) than the South and South-West part of the basin. The combinations of steep slopes and higher elevation may imply a higher chance of susceptibility for land degradation and soil erosion, among other things, in the catchments in this part of the basin resulting in higher percentage of stony upper soil. As can be seen from Figure 2-12, the existing agricultural land coverage is nearly comparable in magnitude to the amount of the available suitable lands for agriculture. Out of the cultivated land in the Jedeb catchment (about 71% of the total area), 8.9% were unsuitable. On the other hand, 22.2% were suitable, but uncultivated. The suitable but uncultivated area, however, may be due to areas that are suitable but protected (such as church compounds and communal grasslands), and thus may be practically unavailable for agriculture.

2.6   Conclusion and recommendations

Results of the land suitability analysis show that suitable and moderately suitable agricultural lands remain available in the Abbay basin, yet marginal and unsuitable lands are being used for agriculture leading to land degradation and soil erosion: a paradox for policy makers. Only a proactive land and water resources governance and policy supported with open data and computing resources can reverse these trends for a more sustainable livelihood. Furthermore, the study demonstrated the use of large sets of freely available global data layers from a number of data sources to compute general land suitability assessment. Such datasets provide alternative access to expensive and proprietary systems for generating information on natural resources which can be especially in developing regions such as in Africa. However, results from such an analysis should only be taken as a preliminary suitability overview of the basin; they need to be verified using ground data and local knowledge before it may be used for decision making. Moreover, the usability of the resulting suitability maps presented in this study in terms of serving as a decision support tool is somehow limited due to its technical complexity as well as reliance on proprietary and expensive analysis/software tools. A web-based framework to automate the gathering and analysis as well as visualization of land suitability mapping may rather be helpful for decision support and overview be it at operational or/and policy levels. The next chapter investigates the development and testing of an automated web-based framework proposed for integrated land suitability assessment in the basin (Yalew et al., 2016a).

Chapter 3. A web-based framework for land-use suitability assessment[5]

3.1   1. Introduction

Massive volume of environmental data is being produced by various global data sources. These data originate from regional and global projects as well as models and satellite imageries (LANDSAT, MODIS) including data on climate change, agro-ecological zones, land cover, terrain, soil, atmospheric and other socio-environmental variables. Managing, analyzing and making meaning out of the overwhelming amount of global data, often referred to as the 'Big Data problem', in a way that supports integrated natural resources management (INRM) remains a challenge (McAfee and Brynjolfsson, 2012; Nativi et al., 2015). Web-accessible computing services are among the best available technologies to overcome this challenge (Vitolo et al., 2015). The Google Earth Engine (GEE), a recently launched platform released for 'trusted testers and partners' as of this writing, presented several features for accessing, computing and visualizing huge data from various data sources to support local, regional and global environmental studies (Gorelick et al., 2017; Moore and Hansen, 2011) .

Spatial decision support systems (SDSS) with implementation of various forms of multicriteria decision analysis (MCDA) methods have been proposed for environmental decision problems whose outcome depends on multiple factors. Land use suitability evaluation is among such environmental problems whose analysis often involves a number of complex and interrelated factors. GIS and spatial analysis tools and techniques are combined with MCDA methods for delivering a better spatial decision by integrating multiple criteria from various spatial data sources. SDSS can be defined as an interactive, computer-based system designed to support decision makers achieve higher effectiveness while solving a spatial decision problem (Malczewski, 2006). GIS-based multicriteria decision analysis (GIS-MCDA) is a class of SDSS that transforms and combines geographic data (input maps) and decision maker'/expert's knowledge and preferences into a decision (output) map (Malczewski and Rinner, 2015). The GIS-MCDA method employs the use of MCDA in the context of spatial decision problems coupled with GIS for enhanced and often spatially explicit decision making (Chakhar and Mousseau, 2008). This method has been widely

---

[5] This material is published as: Yalew, S.G., van Griensven, A., van der Zaag, P, (2016). AgriSuit: A web-based GIS-MCDA framework for agricultural land suitability assessment. *Computers and Electronics in Agriculture* Vol. 128, Pages 1-8. Oct. 2016.

applied for analyzing land suitability for agriculture and/or site selection for various other purposes (Malczewski, 2006; Zolekar and Bhagat, 2015). In land suitability analysis, GIS-MCDA has the advantage of providing structured and spatially explicit evaluation framework for large number of criteria layers.

According to Malczewski and Rinner (2015) , the fundamental procedure for MCDA in general and GIS-MCDA in particular for tackling spatial multicriteria problems involve three main concepts: value scaling (or standardization), criterion weighting, and combination (decision) rule. Value scaling denotes the requirement of transforming the evaluation criteria to comparable units for GIS-MCDA analysis. Criterion weighting involves assignment of weight to an evaluation criterion that indicates its importance relative to the other criteria under consideration. Combination (decision) rule denotes the method of evaluating (and ordering) a set of decision alternatives. A combination rule integrates the data and information about alternatives (criterion maps) and decision maker's preferences (criterion weights) into an overall assessment of the alternatives. A number of approaches and algorithms have been suggested in literature for each of these fundamental procedures (Malczewski, 2006). For criterion weighting, for instance, a vast majority of the GIS-MCDA applications have used one of the three global weighting methods: ranking, rating, and pair-wise comparison (Malczewski, 2006).

The analytic hierarchy process (AHP) is a commonly used MCDA method for determining criteria weights (Saaty, 1988). It employs a pair-wise comparison method which is one of the most widely used procedures for estimating criterion weights in GIS-MCDA applications (Malczewski, 2006). This method is part of the multicriteria decision module in several GIS-MCDA applications (Boroushaki and Malczewski, 2008; Ozturk and Batuk, 2011). AHP simplifies decision making by reducing complex decision problems often involving conflicting factors into pair-wise comparisons. It then derives ratio scales and measures of inconsistency (consistency ratio) from pair-wise comparison of factors and expert judgments. The measure of inconsistency (consistency ratio) was introduced by Saaty (1988) so as to reduce the disadvantage of subjectivity that may be incurred during expert judgments. Combined GIS-MCDA and AHP methods have successfully been applied for land evaluation for various purposes (Anane et al., 2012; Khan and Samadder, 2015). Although several MCDA methods exist including weighted linear combination, ideal point, and out ranking methods (Malczewski and Rinner, 2015), the AHP method was chosen in this development because of the fact that it provides a methodological framework within which inconsistencies in judging the relative importance of factors/criteria in the suitability assessments analysis can be detected and corrected.

The concept of SDSS, and thus GIS-MCDA, has been criticized for the failure to provide suitable tools for wider access for public participation on spatial decisions and for the closed and inflexible nature of available spatial analysis software tools (Malczewski and Rinner, 2015; Sieber, 2006). As can be reflected by the growing interest in development of web-based SDSS (Silva et al., 2014; Terribile et al., 2015), the GIS community seems to be trying to address this criticism by offering analytical and decision support tools that are accessible online to experts and non-experts alike. Besides improving public participation in spatial decision making supported via the internet and web technologies, web-based SDSS also facilitates easier access for diverse spatial datasets from various data sources through a client/server environment. Further development of web-based SDSS may be anticipated to accelerate encouraged by emerging web-based and open-source spatial analysis platforms such as GEE with access for tools and techniques including inbuilt algorithms, spatial datasets and computing capabilities.

In this chapter, we presented the development of a web-based framework (AgriSuit) that can be used for integrated natural resources management in general and land-suitability evaluation in particular. The framework enables the gathering, training and classifying of land cover data as well as assessing land suitability based on GIS, remote sensing and AHP techniques on the Google Earth Engine (GEE) environment. Besides computing and visualizing agricultural land-use suitability, the framework allows for incorporation of ground-based training data and the selection of various algorithms for land-cover classifications. The novelty of this study lies in the development of a framework for integration of globally accessible spatial datasets, algorithms and computing platforms using newly developed and interactive web-tools for regional environmental assessment purposes. Application of the tool on the Abbay (Upper Blue Nile) in Ethiopia basin is demonstrated.

3.2   Data and Methods

Conceptual Framework

The conceptual framework of AgriSuit, shown in Figure 3-1, involves a web-client as the front-end and GEE and QGIS as computing back-ends, with storage and pre-processing tools such as Google Fusion Table (GFT) and webAHP as intermediates. The front-end is the graphical user interface (GUI) web client where users can choose and compute/visualize what kind of suitability (general vs. specific) and based on what strategy and for which year they prefer to analyze.

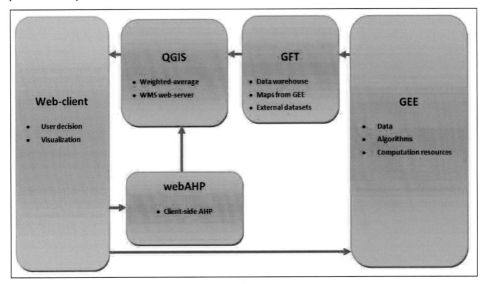

Figure 3-1. Conceptual framework of AgriSuit

The back-end uses GEE for accessing satellite data, land-cover training and classification algorithms as well as Google's cloud computing capabilities. GFT is used to store computed map layers from GEE and other external sources. QGIS is used to compute weighted averages of layers acquired from GFT. Training data are loaded to Google's fusion table which then can be accessed directly from GEE. What is computed from GEE is automatically exported to GFT for storage. Data accessed or computed outside of GEE, such as road and/or river networks as well as other scenario data, are loaded to GFT. GFT and QGIS are connected with a simple JavaScript code for loading maps from GFT to QGIS for geospatial processing. Then, QGIS layers are saved to the QGIS server as Web Map Service (WMS) layers for direct access via the front-end web application.

Data

From literature, we identified important indicators (Table 3-1) including land cover, slope, elevation, soil depth, soil water stress and proximity to various infrastructures and resources (towns, roads, and waters) which determine agricultural land suitability (Mustak et al., 2015; Zolekar and Bhagat, 2015). The GIS criteria layers used for this study are shown in Figure 3-2.

Table 3-1. Data and data sources

River/water bodies	MoWR, Ethiopia	Perennial streams
Towns	CSA (2007), FAO	Woreda towns
Soil group	FAO (FGGD) (2013)	5 arc minute
Slope	Computed from SRTM	30m
Roads	CSA (2007), FAO	All weather roads
Land use	Classified using Landsat 5 (TM) in GEE	30m
Elevation	SRTM (GEE)	30m
Soil Water Content	ISRIC-worldgrid1km (2014)	1km
Soil stoniness	ISRIC-worldgrid1km (2014)	1km
Soil depth	FAO (2014)	5 arc minute

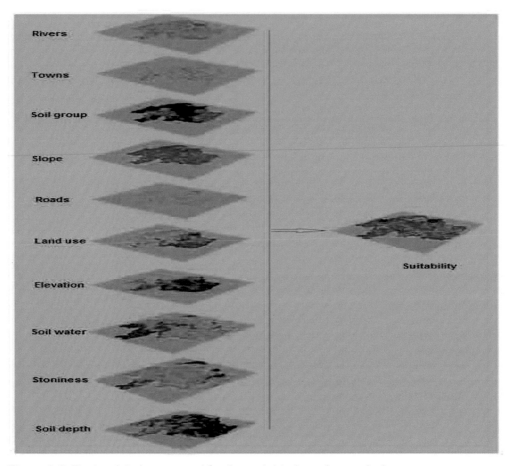

Figure 3-2. Raster data layers used for the weighted overlay analysis

Tools and techniques

Based on FAO guidelines (FAO, 1985; Rossiter, 1996), spatial data for each of the indicator were reclassified into categories of suitability: highly suitable (S1), moderately suitable (S2), marginally suitable (S3) and unsuitable (N). As shown in Figure 3-2, the different GIS criteria layers are then combined to produce a single spatially explicit suitability map that shows the suitability categories of land suitability for agriculture.

Existing as well as newly developed scripts are used for retrieving data, computing, storing and visualizing purposes. Figure 3-3 shows details of execution sequences between the various interacting components of AgriSuit using a sequence diagram. 'Web-client' and 'webAHP' are newly scripted HTML5 and JavaScript based tools, respectively, whereas

QGIS, GFT and GEE are existing tools with newly established link between QGIS and GFT using OpenGIS Reference (OGR) virtual format script (Figure 3-4) and existing link between GFT and GEE for data exchange. Furthermore, land-cover classification and evaluation scripts are newly developed based on algorithms and tools available on the GEE platform. Description of the newly developed scripts and the GEE platform with respect to the sequence diagram in Figure 3-3 is discussed in the following subsections.

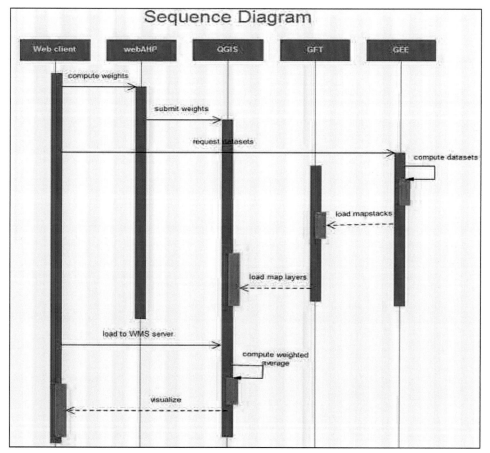

Figure 3-3. Sequence diagram of AgriSuit execution

```
<OGRVRTDataSource>

  <OGRVRTLayer name="Abbay_basin">

    <SrcDataSource>

      GFT: email=email password=password

    </SrcDataSource>

    <SrcLayer>1AkEaPy8qIQCvIDzd7mXiRmGRUKWDM5RnKeWFzhOW</SrcLayer>

  </OGRVRTLayer>

</OGRVRTDataSource>
```

Figure 3-4. OGR virtual format connecting QGIS and GFT. Note that 'email' should be Google authorized Gmail account.

3.2.1.1 Web-client

This is the web-based graphical user interface (GUI) that interacts with users of the system. It contains descriptions of methodologies used for the dynamic analysis and available choices for computation and visualization of land suitability. As shown in Figure 3-3, the web-client invokes the webAHP for computation of weights, it initiates data request from GEE and calls WMS server on QGIS. Since all storages and computing operations are made on the cloud (GFT, GEE), the web-client can be accessed from any browser supporting device such as mobile, laptop or desktop computing devices. However, the framework is developed mainly with PC environments in mind and thus it is not optimized for mobile applications.

3.2.1.2 webAHP

Weights or level of influences of each raster layer on agricultural land-use suitability were computed using the Analytical Hierarchy Process (AHP) technique. AHP is a structured and organized method for analyzing complex decisions by breaking them into pair-wise alternatives (Saaty, 1988). For the AgriSuit framework, a client-side AHP algorithm (webAHP) was developed using JavaScript. After the weight of each criteria layer was computed using webAHP, a GIS based weighted overlay (weighted average) analysis was carried out to produce a single suitability map (see Figure 3-2).

The webAHP interface prompts users to enter criteria layers. Then users are prompted to enter their ratings about how important each of the defined criteria are for determining agricultural land-use suitability in a pair-wise comparison (Figure 3-5). For instance, if ten criteria are defined for determining land-use suitability, each of the ten criteria is iteratively presented paired with another for expert/user comparison. Computation of consistency ratio (CR), which is a measure of how consistent the judgment on criteria comparison have been relative to large samples of purely random judgments (Saaty, 1988), is also implemented in this tool to detect and correct inconsistencies during expert judgments.

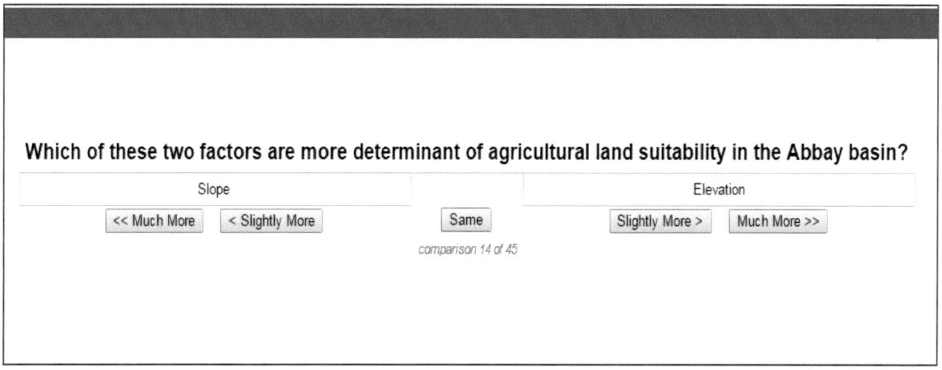

Which of these two factors are more determinant of agricultural land suitability in the Abbay basin?

Slope			Elevation	
<< Much More	< Slightly More	Same	Slightly More >	Much More >>

comparison 14 of 45

Figure 3-5. Pair-wise comparison of suitability criteria on webAHP

At the end of the pair-wise evaluation exercise for all the criteria combinations using webAHP, weights are computed for each criteria layer and passed to QGIS for assessing the overall weighted average suitability layer. Note that we selected 10 criteria layers based on literature for the suitability assessment in this study. However, webAHP and the associated Web-client can interactively take more or less criteria layers as per user specifications.

3.2.1.3 GEE

GEE is a platform designed to enable planetary-scale scientific analysis and visualization of geospatial data. Currently released only for 'partners, developers and trusted testers', it is a platform that makes available nearly 40 years of the world's satellite imagery with cloud computing resources and tools for scientists and researchers[6]. The platform provides computational power using Google's parallel processing power and with access to develop and/or run algorithms on the full Earth Engine data archive. Its applications include detecting deforestation, classifying land-cover and estimating forest biomass and carbon (Moore and Hansen, 2011). In this framework, GEE is used for gathering and processing satellite imagery

---

[6] https://earthengine.google.org/

data and for training and classifying land-cover. Since GEE currently does not allow direct connection of third-party applications and tools with it, GFT is used here as an intermediary access and storage point for computed resources from both GEE and external applications. The connection between GFT and QGIS, as shown in Figure 3-4, is established using a simple OGR virtual format script.

For the application in the Abbay basin, training point data from a dozen of known locations for each land cover types were identified on Google Earth and exported as *.kml file format to GFT. Then, these training points were loaded from GFT to GEE using JavaScript codes (Figure 6). Afterwards, Landsat collections (images) were loaded for training and classification. For this case study, land-cover classification was processed from 1984 to 2015. A number of classifier algorithms including Support Vector Machine (Peng et al., 2002), Classification And Regression Tree (Bel et al., 2009) and Fast Naive Bayes Classifiers (Dietterich, 1998) are available inbuilt on GEE.

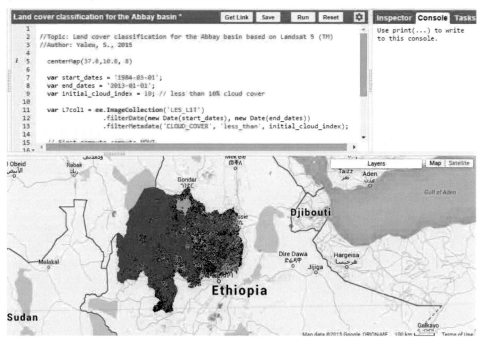

Figure 3-6. GEE coding and visualization interface

3.3   Results and discussion

The overall output of the framework is a simplified web interface where users can compute weights using the AHP algorithm (webAHP), and analyze and visualize land suitability for agriculture. Users may also animate trends of suitability changes over time. In case new land-use training data is made available, users have the option to upload the training data, choose year of the training data, choose classification algorithms, and classify land-use that will then be used for the suitability computation. The weights of the different criteria layers computed using webAHP for the Abbay basin application are shown in Figure 3-7. According to the webAHP computation result, slope has the largest weight for determination of land suitability in the study area with a value of 0.19 followed by soil stoniness and soil moisture with weight values of 0.17 and 0.13, respectively. On the other hand, elevation has the smallest weight value of 0.04, followed by soil type and land use\land cover with values of 0.04 and 0.05, respectively. A horizontal bar chart on the right hand side of each criteria factor shown in Figure 7 allows for a quicker visual comparison of the criteria weights. A consistency ratio value of 0.05 is achieved in this application as shown at the bottom end of Figure 3-7. The Weights can be scaled or re-scaled based on user preference for a better visualization. The Web-client (Figure 3-2) presents visualizations of choices and results that include maps, customized legends, graphs and explanation of data, methods, guidelines as well as resources and theoretical backgrounds used for the development of this framework. On the left panel of the Web-client interface, users can chose types of suitability (specific vs. general), select or introduce strategies (policy constraints or incentives on the various criteria factors), and choose years for which the suitability is computed. As shown on the same panel in the figure, users can switch between different case study areas in the 'Change catchment' dropdown box. Case study sites are automatically detected from the QGIS data store (WMS web server) and dynamically added to this list at run time. Besides the map visualization in the center panel at flexible zoom levels, graphs are presented on the right panel of the interface for quick assessment. Land cover training data can be uploaded either as vector, raster, json, or kml file formats on the bottom panel. Based on uploaded training data, users can select training land cover data year and training algorithm for online reclassification of a new land cover map for use to a suitability computation.

Result : Which of these two factors are more determinant of agricultural land suitability in the Abbay basin?

Option	Result	Scaled Result	
Land cover	0.0471	0.0471	
Slope	0.1922	0.1922	
Soil stoniness	0.1657	0.1657	
Soil type	0.0411	0.0411	
Soil depth	0.1123	0.1123	
Soil moisture	0.1255	0.1255	
Elevation	0.0378	0.0378	
Distance from town	0.1077	0.1077	
Distance from roads	0.1018	0.1018	
Distance from water	0.0689	0.0689	

1 | Scale

Consistency Ratio: 0.0500

Retry | Hide/Show

Figure 3-7. Weighted averages (suitability weights) computed using webAHP

The AgriSuit framework therefore reduces the often daunting task of land-cover classification into a 'click and play' task. Unlike the traditional suitability analysis practice which often involves desktop GIS preprocessing of criteria layers as well as the use of static land use/land cover maps derived once from historical data, this framework provides an interactive and dynamic environment for input preprocessing as well as computations of resulting maps (Figure 3-8).

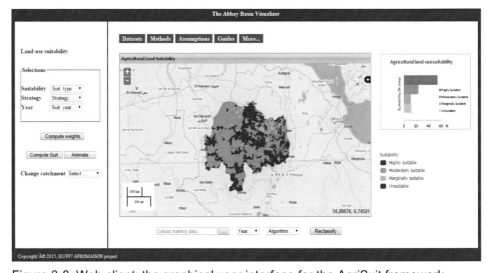

Figure 3-8. Web-client: the graphical user interface for the AgriSuit framework

AgriSuit can be an important tool both at operational as well as policy level overviews of land suitability assessment. At operational level of this demonstration application, for instance, local farming extension workers in the Abbay basin can find it an important spatial decision support tool as their task involves advisory service and recommendation to farmers of agricultural land suitability for various purposes. At policy level, regional land and watershed decision makers can be better informed using this spatial decision support framework for a spatially explicit overview of locations and extents of various categories of general agricultural land suitability.

AgriSuit and the GEE computing environment it operates on present a number of opportunities compared to the traditional desktop based and manual computation of environmental data for similar purposes. First, it gets unlimited access to stores of global datasets from GEE. This helps evaluate alternative data from various sources. It also simplifies time intensive manual data gathering and pre-processing to a trivial task of clicking and applying. Second, a number of inbuilt algorithms for data training, classification and evaluation purposes are easily available through the use of GEE. This brings a no-cost alternative for proprietary and highly expensive (especially for developing regions) desktop spatial analysis software tools. Third, the processing is based on Google's cloud computing platform. This means that time intensive remote sensing computations such as for data training and for land-cover classification can be computed in a matter of minutes irrespective of the spatial coverage of the analysis. Lastly, AgriSuit as well as the tools used for developing this framework are free and open source. This enables people to develop, modify or customize any of the tools and algorithms for their own local cases. Open environmental data and computing resources are creating a level playing field for environmental modelers and decision makers, particularly for those from least developed regions. The increasing availability of such open and high resolution global data and computing resources has reduced the digital divide in sustainable management of regional environment.

Two aspects of limitations may be highlighted from this chapter: the first limitation is pertaining to the presented framework itself and another pertaining to the demonstrated case study application. Limitations with the framework mainly emanate from the difficulty of addressing inconsistency during, for instance, pair-wise comparison of criteria for weight computations using the AHP technique. The choice of 'slightly more', 'same' and 'much more' as bases for weight computations is a difficult limitation that introduces inconsistency inherent to the use of AHP techniques. To reduce the impact of this limitation, the consistency ratio (CR) measure is implemented in webAHP. According to Saaty (1988), if the CR is much in excess of 0.1 the judgments are untrustworthy because they are too close for randomness. By this measure, the CR value of 0.05 computed in this study (Figure 3-7) is logically consistent enough for use in the suitability assessment application procedure. Furthermore, the framework assumes a certain level of understanding of land cover training and classification concepts by end users for incorporating new land cover maps for suitability computation. Even though feedback is implemented showing land cover classification performance using Kappa indices (Pontius, 2000), the overall quality of the land cover produced in this way and henceforth the land cover factor for land suitability assessment may be as good as the technical level of the end users. The second limitation is that this framework is currently developed with PC screens in mind and thus not optimized for smaller mobile screens such as cell phones. This might be a drawback especially for operational use in its current implementation. Limitations pertaining to the demonstrated application in the Abbay basin are related to data quality. Obtaining local and/or regional datasets for input to the suitability computation in the study area is challenging. The suitability results presented in the demonstration are based mainly on global data sources without field verification. Therefore, any quantitative results associated with the suitability categories from this application should be interpreted with caution.

Despite these limitations, the framework can easily be used to compute land suitability from globally available data sources at real-time, the results of which can directly be used for overview of regional and local land suitability assessment for environmental planning purposes. Furthermore, the framework enables users to upload or replace global datasets for any of the criteria layers when local or higher quality dataset becomes available.

3.4   Conclusion

This chapter presented a framework for land-use suitability analysis using web-based GIS-MCDA methods. It provides a web-based graphical user interface that provides an easier access to spatial datasets, algorithms and computing capabilities based on the GEE platform. The framework enables the gathering, pre-processing, computing and visualizing of data and criteria layers for land suitability evaluation. Criteria layers are combined to produce a suitability map through weights derived using a client-side AHP tool developed and integrated in the AgriSuit framework. The framework presents a number of opportunities for potential land suitability assessment endeavors including easier access for data, algorithms and computing resources. Furthermore, this framework as well as the tools it operates on are open-source and freely accessible. Limitations of this study include those that are pertinent to the framework itself such as inherent issues with the use of AHP methods (inconsistencies with expert judgments) as well as pertinent to the demonstrated application in particular (such as data quality). For future studies, the AgriSuit framework can be optimized for mobile devices. Furthermore, besides AHP, additional options of MCDA including weighted linear combination, ideal point, and out ranking methods can be tested. In the next chapter, we will develop and validate a land-use change model for the Abbay basin based on the suitability analysis methods discussed thus far and socio-environmental land-use change drivers for allocation of land-use for various purposes.

Chapter 4. Land-use change modeling for the Abbay basin[7]

4.1 Introduction

Current rates, extents and intensities of land-use and land-cover change are driving unprecedented changes in ecosystems and environmental processes at local, regional and global scales. As a result, environmental concerns including climate change, biodiversity loss, land-degradation, soil erosion and pollution of water and air are growing. Interaction of the changes in land-use and land cover with various subsystems of the earth system including hydrology, the climate system, biogeochemical cycling, ecological complexity and land degradation make the study of this subject a complex science (Fürst et al., 2013; Halmy et al., 2015; Rindfuss et al., 2008; Turner et al., 1995). Monitoring and mediating the negative consequences of land-use and land-cover change while sustaining the production of essential resources has therefore become a major priority of researchers and policymakers around the world (Ellis and Pontius, 2007). However, analysing the fundamental socio-political, economic, cultural and biophysical forces that may drive land-use and land-cover dynamics and predicting a likely trajectory of future changes constitute one of the main challenges in land-use research (Geist and Lambin, 2002; Rendana et al., 2015; Veldkamp and Lambin, 2001). Land-use modeling is often used for predicting trajectories of future landscapes. A typical approach to land-use change modelling involves investigating how different variables relate to historical land-cover change trends and transitions in the past and use those relationships to build models that project a likely future land-use trajectory (Chen and Pontius Jr, 2010; Pontius Jr et al., 2008).

The Upper Blue Nile (Abbay), despite being one of the most diverse and highly important river basins in Ethiopia, faces serious problems including soil erosion, land degradation, loss of soil fertility and deforestation (Asres, 2016; Urgesa et al., 2016). The major causes are reported to have been a combination of biophysical factors such as seasonal fluctuation in rainfall and climate variability, topographic heterogeneities, and anthropogenic factors, e.g. population growth and associated demands, that result in soil erosion and land degradation in the basin (Bewket and Sterk, 2002; Hurni et al., 2005; Setegn et al., 2009; Steenhuis et al.,

---

[7] This chapter is based on a journal paper published on the journal of Environments:

Yalew, S.G. , Mul, M.L., Teferi, E., van Griensven, A., Priess, J., Schweitzer, C., van der Zaag, P. (2016) Land-use Change Model for the Upper Blue Nile Basin. Environments 3.3 (2016): 21.

2009). Land degradation occurs mainly due to gully and surface erosions by torrential runoff in this rugged highland catchment. No predictive land-use change modelling study addressing socio-economic and biophysical land-use change drivers has yet been reported in the Abbay basin in general and in the Jedeb catchment in particular.

Several land-use change modelling tools have been developed in the past (Brown et al., 2004). The models differ on the scale of application; whether they are deterministic or probabilistic; and whether they are spatially explicit or spatially inexplicit. Spatially explicit models show where and how much land-use change is occurring with implications of why change is occurring. Spatially inexplicit models show only cumulative changes of land use irrespective of where the change takes place. Model accessibility is also another differentiating factor. Some land-use change models have open source access policy, e.g. SITE (Simulation of Terrestrial Environments) (Schweitzer et al., 2011) and GAMA (Amouroux et al., 2009), whereas some others are proprietary, e.g. GEONAMICA (Hurkens et al., 2008). In addition, some land-use models focus only on biophysical or only on socio-economic driving forces of land-use change whereas others try to combine both. Land-use change models vary in complexity and flexibility as well. For this study we have chosen to use the SITE (SImulation of Terrestrial Environment) land-use modelling framework due to its suitability for representation of socio-economic as well as biophysical inputs and for its capability of spatially explicit land-use change simulation. SITE is a cellular automata based multi-criteria decision analysis framework for simulating land-use conversion based on socio-economic and environmental factors (Schweitzer et al., 2011). It also provides a number of algorithms and tools such as for model evaluation, calibration and visualization. In addition, the model can be easily modified as it allows access to its source codes of the underlying modelling sequence. Model evaluation and calibration has been depicted as one of the challenging tasks in land-use change modelling due mainly to the level of complexity the subject presents (Herold et al., 2005; Wegener, 2004).

This chapter is aimed at identifying the potential land-use drivers in the Jedeb catchment of the Abbay basin by combining statistical analysis, field investigation and remote sensing. Potential future trajectory of land-use change was predicted under a business-as-usual scenario in order to provide critical information to land-use planners and policy makers for a more effective and proactive management in this highland catchment. To do so, a land-use change model was setup, calibrated and evaluated using the SITE modelling framework. Note that land cover is the observed biophysical cover on the earth's surface whereas land use is characterized by activities and inputs people undertake on land cover type to produce, change or maintain it (Di Gregorio, 2005). In this chapter, we are simulating changes in land cover using land-use drivers as well as baseline and reference land-cover maps.

4.2   Materials and methods

Study Area

The Jedeb catchment is situated in the south-west part of Mount Choke and it is part of the headwaters of the Abbay basin (Figure 4-1). It covers an area of 297 km2 and lies between $10°22'$ to $10°40'$ N and $37°33'$ to $37°50'$ E. The area is known for its diverse topography with elevation extending from 2,100 to 4,000 m.a.s.l., and slope ranging from nearly flat to very steep (> 45°). The mean annual rainfall varies between 1,400 and 1,600 mm/a (based on data from 3 climate stations: Debre Markos, Anjeni and Rob Gebeya). The steep slopes, coupled with erosive rains, have contributed to the excessively high rates of land degradation and soil erosion (Betrie et al., 2011). As one of the severely eroded and degraded parts of the basin, the catchment received the attention of researchers who undertake various socio-environmental and water resources studies in the catchment (Teferi et al., 2013a; Tesfaye and Brouwer, 2012). Land-use and land cover changes, such as loss of grassland cover due to overgrazing, poor land-use management, and change from grassland to agricultural land, for instance, may have contributed to a higher level of gully formation, soil erosion and land degradation. Between 1957 and 2009, 46% of the watershed has undergone land-use changes without proper soil and water conservation measures in place (Teferi et al., 2013a). The changes in land-use and land cover are thought to be among the major causes of high erosion rates in the basin (Bewket and Teferi, 2009; Pankhurst, 2010). Whether this trend will continue is dependent, among other things, on future land use.

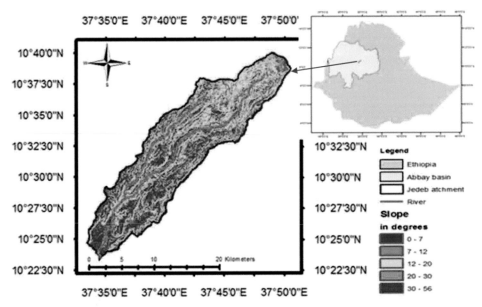

Figure 4-1. Location and topographic map of the Jedeb catchment in the Abbay basin, Ethiopia

Conceptual framework

First, detailed land cover maps for the years 1986 and 2009 derived from Landsat TM Satellite images were used as base and reference (hence forth 'observed') maps for the land-use model, respectively. Then land-use change drivers were identified and the strength of their influence on land-use change estimated by analyzing the spatial correlation between an initial set of potential drivers and land-use types. Rule-sets and initial weights for each deriver variables were developed for each land-use type based on the correlation results. Then, a spatially explicit land-use change model was developed on SITE using the identified land-use drivers and the 1986 land cover map. Based on land-use suitability factors and historical demands for various land-use types (section 4.2.3.4), dynamics of trends of land-use conversion was simulated and analyzed between 1986 and 2009. The simulated output map of 2009 was compared with the reference or observed land cover map of 2009. The model was then calibrated based on field data, trend analysis, and secondary data sources. The model was thereafter used to simulate a business-as-usual scenario of land-use change trajectories of the catchment for the year 2025.

Inputs and model setup

4.2.1.1 Model structure

SITE (Simulation of Terrestrial Environments) is a generic and spatially explicit land-use modelling framework based on an extended cellular automata and multi-criteria concept (Mimler and Priess, 2008). It employs a rule-based approach for assessing land-use suitability based on various criteria: ecological, economic, cultural and demographic factors as well as neighbourhood effects. It simulates land-use dynamics in an annual time step. It is also an open source, flexible and extendible land-use modelling framework. Taking in to account socio-economic as well as biophysical aspects, it has a capability of simulating land-use suitability, dynamic land-use changes, vulnerabilities and potential consequences of various land-use management measures. Simulations are carried out in the following sequence:

1. Multi-criteria suitability assessment: including ecological, economic and cultural and demographic factors as well as neighbourhood effects (spatial auto-correlation).
2. Decision making based on the suitability assessment as well as regional constraints, rules, regulations & regional preferences.
3. Land allocation driven by demands for spatially relevant commodities.
4. Calculations of ecosystem services and land-use related changes in biodiversity.

The framework has been used to assess socio-environmental and land-use dynamics in the context of natural resources management on case studies in various parts of the globe. It has been applied, among others, in Indonesia to simulate land focusing on socio-economic and environmental effects of different strategies of resource use (Priess et al., 2007); in India to analyse trade-offs of land-use change with regard to the production of bio-energy (Das et al., 2012); in Mongolia to study regional land dynamics with a strong focus on the linked impacts on water resources (Priess et al., 2011; Schweitzer et al., 2011), and in Ethiopia to simulate land-use suitability based on multiple socio-environmental factors (Yalew et al., 2012) .

SITE consists of two main components: (i) the system domain (SD) which includes optimized methods, procedures and essential tools for the modelling process, implemented in C++, and (ii) the application domain (AD), which is a python interface designed for the development of applications and decision rules to address case specific implementations (Figure 4-2). Based on suitability and driven by demands for spatially relevant commodities, the model allocates land for defined land-use classes. Its modular implementation and source code accessibility (open source) makes extending the application or coupling third-party software relatively easy. SITE includes modules for calculating suitability and for allocating land-use classes based on suitability.

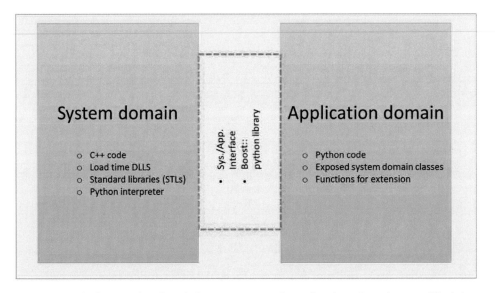

Figure 4-2 Software details of the system and application domain; modified from Mimler and Priess (2008)

Land-use suitability is calculated within the suitability module of SITE. This module is subdivided into functions computing biophysical suitability (e.g. elevation, terrain slope, soil fertility) and socio-economic suitability (based on factors, such as population, gross margins, accessibility and farmers' preferences) to produce land use suitability maps (Schweitzer et al., 2011). All suitability values in SITE are normalized to a range between 0 (not suitable) and 1 (perfectly suitable) using the following Eq. 3.1 (Mimler and Priess, 2008).

$$S_{kl} = (W_B \sum_{i=1}^{m} \beta_i S_{Bikl} + W_E \sum_{i=1}^{n} \varepsilon_i S_{Eikl}) * \prod_{j=1}^{o} C_{Bjkl} \prod_{j=1}^{p} C_{Ejkl} \qquad (3\text{-}1)$$

$where\ W_B + W_E = 1;\ \sum_{i=1}^{m} \beta_i = 1;\ \sum_{i=1}^{m} \varepsilon_i = 1\ and\ S_{Bikl}, S_{Eikl}, C_{Bjkl}, C_{Ejkl} \in [0,1]$

The calculation of the overall suitability value S_{kl} for each land use grid cell k and land-use class l consists of two terms: the partial suitability S_{Bikl} for biophysical and S_{Eikl} for socio-economic factors. The C_B and C_E variables are biophysical and socio-economic constraints, respectively. These factors are weighted using the partial weights β_i / ε_i, where m and n

represent the total number of suitability criteria included. Variables o and p represent the total number of biophysical and socio-economic constraints, respectively.

Suitability of land use is, thus, defined in this implementation by analyzing spatial correlations of where a specific land-use is found with respect to factors such as slope, elevation, soil, etc., as well as by historically established links between land use and various socio-economic aspects. Each land-use class, thus, is spatially correlated with a group of attribute sets driving its conversion (such as slope, elevation, and proximities to water, road and markets). The allocation module of SITE uses a suitability map, set of neighborhood functions and defined socio-environmental factors, for allocating land-use types. It follows defined hierarchical priorities and land-use change rules. Suitability factors show what combination of major criteria are suitable for which land uses and hierarchical priorities show which land-use type takes priority during allocation in case a land parcel is suitable for two or more competing land uses.

4.2.1.2 Land-use change drivers

Potential land-use change drivers were gathered through literature review and interview with key informants including farmers, regional and local land resources administrators and development agents (i.e., government employees assigned in villages to advise farmers on various agro-ecosystems practices and local and regional land administration policies). Land-use practices and perceptions of farmers on issues such as availability of land for various land uses, perceived changes in the past and their anticipation of future prospects with regards to land use, their practice of crop-rotation and trends and traditions of land-renting were reflected. What the farmers consider as limiting factors of productivity such as access to water for irrigation, roads for transport of products, drought/rainfall limitation, and lack of agricultural and grazing land were also deliberated. In addition, regional and national land-use policies, and national growth and development plans were consulted. Suitability relevance of distance variables from such as urban centers, water bodies and roads were estimated based on literature and discussions with local experts. The outcome of the discussion with local experts and stakeholders is mainly qualitative, yet it served as a basis for parameter estimation, in addition to relevant literature, of initial suitability ranges.

Data reduction and correlation analysis between the identified potential drivers and land-uses were conducted using the Principal Component Analysis (PCA) method (Abdi, 2003; Abdi and Williams, 2010). PCA produces correlations between variables by identifying hidden patterns in data and classifying them according to how much of the information is stored in the data they account for (Jolliffe, 2005). PCA has been used in literatures to analyze land cover changes and land use change drivers(Du et al., 2014; Skånes and Bunce, 1997). Eleven potential land-use drivers (population, distance to market, distance to road, slope, distance to settlement, elevation, livestock, soil type, precipitation, distance to forest edge), and distance to water sources were identified as input to the PCA analysis. By applying the PCA using these driving factors, land-use change drivers that capture most of the variations in change for each land-use type can be identified. Then, comparative significance (initial suitability weight) for each of the associated land-use change drivers was established using Eq. 3.2. The suitability weights show the importance of each suitability factor in determining the land-use type.

$$\alpha = \left(\beta \Big/ \sum_{i=1}^{y} \beta_i \right) \qquad\qquad (3\text{-}2)$$

where α = comparative significance (initial weight) value; β = individual significance value; y = number of significant independent variables for the land-use class. The quotient of individual significance values and the sum of all the significance values of determinants (land-use drivers) for a land-use class is a normalized value showing an initial weight between 0 and 1 (Note that in the absence of means of estimation of initial weights for suitability factors on the ground, it is a common practice in SITE to set default weight of 1 for each suitability subset. This would, however, mean that besides the need for increased computation time, the model will be forced to 'fit' parameters to past observations during calibration irrespective of relevance on the ground). Assignment of an initial weight for calibration reduces the computation burden in addition to serving as a model evaluation tool comparing weights estimated based on ground data against model calibrated values. Initial weights can be altered during model calibration.

4.2.1.3 Data

Potential land-use change drivers were identified through literature reviews (Serneels and Lambin, 2001; Teferi et al., 2013a; Veldkamp and Lambin, 2001) and field interviews with farmers, local farming experts, regional land bureau officials, and through spatial correlations (Table 4-1). In addition to derived spatial layers such as distances from roads, towns and rivers, a number of biophysical and socio-economic datasets were gathered, pre-processed and used in the land-use model setup (Table 4-1). Major socio-economic data were collected from the Ethiopian Statistical Agency (CSA, 2007), Atlas of Ethiopian Rural Economy (Chamberlin et al., 2007) and the Ethiopian Rural Household Survey(1989-2009) (Dercon and Hoddinott, 2004). Field observations and interviews with key informants also provided valuable insights in the identification of land-use change drivers in the catchment. The base and reference Landsat Thematic Mapper (TM) based land cover maps for the years 1986 and 2009, respectively, were produced from a previous study carried out in the catchment (Teferi et al., 2013a). The land cover classes were reclassified into 5, i.e., Natural Woody Vegetation (NWV), Plantation Forest (PF), Cultivated Land (CL), Grassland (GL), and Others. To shortly summarize the land-cover classification procedure, a hybrid (supervised and unsupervised) classification approach was adapted with successive GIS/spatial operations for classifying the imageries. Multispectral pattern recognition using the Iterative Self-Organizing Data Analysis Technique (ISODATA) algorithm (Ball and Hall, 1965) was performed on the imageries for the land cover classification. Field data was collected to associate the spectral classes with the cover types in the classification scheme for the 2009 Landsat imagery. Reference data for the 1986 image were based on aerial photo interpretation of 1982 as well as topographic maps of 1984 at a scale of 1:50,000 collected from the Ethiopian Mapping Agency (EMA). Of a total of 2277 reference data points for the respective years, 759 points were used for accuracy assessment and 1518 points were used for classification. Training sites were developed from the field reference data collected to generate a signature for each land cover type. An overall accuracy of 95.6% and a Kappa coefficient of 0.94 were attained for the 2009 classified map. Similarly, overall classification accuracy of 91.5% (Kappa coefficient of 0.89) were achieved for the 1986 land-cover map (refer to Teferi et al. (2013) for details on the land-cover classification).

Table 4-1 Data inputs and potential land use change drivers

Variables	Description	Dataset	Sources[*]	Scale/ resolution
Population	Gridded population dataset	Census for 1986 & 2007; GPW	CSA, FAO	Sub-district; 1km
Livestock	Gridded livestock dataset	Gridded livestock (GLW) 2007, 2014	FAO	5km
Distance to roads	Euclidean distance to major roads	Roads	ERA	30m
Distance to markets	Euclidean distance to major towns	Markets	FAO-SRDN	30m
Land cover map	Land cover maps	Landsat TM (1986 & 2009)	Teferi et al., 2013	30m
Settlement maps	Topographic map with settlement locations	Topo1984; Landsat	EMA, GEE	1:50,000; 30m
Crop map	Map of croplands in the Amhara region	Cultivated land	BoA, MoARD	250m
Distance to water	Euclidean distance to water sources	Water bodies	MoWE	30m
Slope and elevation	Elevation (DEM) and slope (derived from DEM)	DEM	USGS	90m
Soil type	Soil types	Soil group	FAO/FGGD (2013)	5 arc min.
Precipitation	Average annual precipitation	Precipitation data	MoWE	Annual average
Distance from forest edge	Distance from forest edge	Distance from forest edge	land-use map	30m

[*] CSA=Central Statistical Agency of Ethiopia; ERA: Ethiopian Roads Authority; EMA: Ethiopian Mapping Agency; FAO=Food and Agriculture Organization; GLW: Gridded Livestock of the world, an FAO project; GPW= Gridded population of the world; GEE= Google Earth Engine; BoA: Amhara Bureau of Agriculture; MoWE: Ministry of Water & Energy of Ethiopia; EMA: Ethiopian Meteorological Agency; USGS: US Geological Survey; MoARD: Ministry of Agriculture and Rural Development.

4.2.1.4 Demand for land use

Land-use change is driven by demands for various uses. The demands are associated with livestock and population and thus can be affected by factors at local, regional as well as global scales. Land-use demands include settlements, food production, and lifestyle needs; fodder and grazing needs; and/or nature protection/conservation needs, etc. If population increases, one may assume that demands for settlement (especially near urban areas) and cultivation or livestock (in the rural lands) may be higher. Based on case specific information, the amount of added population every year needs to be taken into account and allocated for settlement, cultivation and livestock/grazing requirements. In this case study, human population as well as livestock growth rates were taken from regional datasets, specific demands were estimated based on field investigations and findings from the literature review.

Based on field investigation and the literature, minimum requirements for various land-use types in the catchment were estimated per household, Table 4-2. The average number of people in a household is assumed to be the current regional average of 4.3 (CSA, 2008). Socio-economic demands were estimated based on the projection of the regional growth rates for population and livestock. The historical growth rate for population and livestock for the simulation period were 2.5 % and 1.5% per annum, respectively (CSA, 2007; FAO, 2004b). For instance, demand for settlement or cultivation is expressed based on average individual demands (Table 4-2). Likewise, demand for grassland (for instance) is computed based on average livestock demand (Table 4-1). The demand variables are, therefore, expressed in terms of population and livestock in this case and amount is spatially-explicitly computed in the rule-sets/application codes of SITE. Demands for plantation forest estimated per household after field investigation.

Table 4-2. General demand estimations based on Mengistu (2006) and Jayne et al (2003)

Variable	Estimated value
Cultivation requirement	1.17 ha/household
Settlement requirement	0.25 ha/household
Plantation (for fire wood, housing) requirement	0.06 ha/household (about 1/20th of cultivation/household), field survey
Grassland (grazing) requirement	0.25 ha/livestock

4.2.1.5 Model setup

SITE requires a number of GIS based pre-processing tasks of spatial and socio-environmental data. The model inputs include land-use change drivers, are pre-processed in formats required by the tool (Schweitzer et al., 2011). SITE was set-up using various socio-economic and biophysical data (Table 4-1) in this chapter. The land-use model was run on a 200m by 200m grid and a one year time step. The rule sets define the dynamics of the land-use simulation starting from the base year of the simulation (1986).

4.2.1.6 Model evaluation

Initial values for suitability weights and ranges, obtained from the analyses described earlier, were applied to parameterize the model. The initial weight parameters were adjusted through model calibration until a good fit was obtained. The GALib genetic algorithm library (Wall, 1996), which is already embedded in SITE, was used for this purpose. Initial suitability weight parameters for slope, elevation, and distance variables (distances from settlement, roads, market, and forest edge) were subjected to the calibration algorithm. Land-use change model results are often evaluated by comparing simulated maps against a reference map. Similarly here, the simulated raster output of the model for 2009 was evaluated against the reference (Landsat derived) map for the same year. Depending on the data structures of the resulting output (raster, vector, or hybrid), a number of algorithms have been developed over the years for comparing two maps. However, there does not seem to exist any agreed universal procedure to do that (Kuhnert et al., 2005). For a spatially explicit, grid-based categorical data (such as land-use or vegetation classification presented here), cell-by-cell comparison to get the number of matching cells, Eq. 3.3, is often the simplest (Kuhnert et al., 2005; Visser and De Nijs, 2006).

$$C_C = N_M / N_T \qquad\qquad\qquad\qquad (3\text{-}3)$$

where C_C= is coefficient of cell agreement, N_M= number of matched cells, N_T = number of total cells.

Problems with cell-by-cell comparison arise from the fact that if one of the maps is shifted even by a single cell, the agreement of the whole comparison may be compromised. Due to lack of accounting for allocation of the neighborhood cells, a small disagreement and a big disagreement can have the same error value. It was progressively noted that a full characterization of a fit between two maps should tackle not only quantity, and or location of changes of matching cells but also distances between locations of matching cells (Kuhnert et al., 2005). To address this and a number of other map comparison bottlenecks (Hagen-Zanker and Lajoie, 2008; Pontius Jr and Millones, 2011; Pontius, 2004), alternative algorithms have been proposed over the years (Pontius Jr and Millones, 2011; van Vliet et al., 2011; van Vliet et al., 2013). Pontius and Millones (2011) suggested that summarizing cross-tabulation matrix of the simulated and the observed land-use map in terms of quantity and spatial allocation disagreements will sufficiently account for differences between two categorical maps in terms of the quantity (changes or persistence) and allocation of matching cells. A variety of statistical summaries of a cross-tabulation matrix tool has been recommended (Pontius and

Millones, 2011). The cross-tabulation tool provides one comprehensive statistical analysis to answer two important questions simultaneously, that is, how well two maps agree in terms of the quantity of cells in each category and how well they agree in terms of allocation of cells in each category. Eq. 3.4 and Eq. 3.5 represent quantity and allocation disagreements for two categorical maps, respectively (Olmedo et al., 2015; Pontius and Millones, 2011).

$$QD = \frac{\sum |\frac{n_{+i}}{n} - \frac{n_{i+}}{n}|}{2} * 100 \tag{3-4}$$

$$AD = \frac{\sum (2*\min(\frac{n_{+i}}{n} - \frac{n_{ii}}{n}, \frac{n_{i+}}{n} - \frac{n_{ii}}{n}))}{2} * 100 \tag{3-5}$$

where QD=quantity disagreement; AD=allocation disagreement; n_{ii} = diagonal matrix elements; n = total number of considered pixels; and n_{+i} and n_{i+} = marginal sum of the columns and marginal sum of the rows in the error matrix, respectively.

Quantity disagreement is the difference between two maps due to an imperfect match in overall proportions of all mapped land-use categories whereas allocation disagreement is the difference between two maps due to an imperfect match between the spatial allocation of all mapped land-use categories (Pontius and Millones, 2011). Values from comparison of two maps using this measurement technique range between 1 (100%) (Perfect disagreement) and 0 (0%) (Perfect agreement). Interpretation of what is good level of agreement in map comparisons is rather subjective. Landis and Koch (1977b) lumped possible ranges of map comparison into three groups: agreement value greater than 0.8 (80%) represents strong agreement; agreement value between 0.4 (40%) and 0.8 (80%) represent moderate agreement; and agreement value less than 0.4 (40%) represents poor agreement between two maps. An interpretation by Altman (1991) states that comparison agreements are 'very good' if two maps agree by more than 0.8 (80%); 'good' if they agree 0.6 (60%) to 0.8 (80%); 'moderate' if they agree between 0.2 (20%) and 0.6 (60%); and poor' if they agree by less than 0.2 (20%). In this chapter, the simulated land-use maps were evaluated against the reference map using the Quantity and Allocation Disagreement measures (Olmedo et al., 2015; Pontius Jr and Millones, 2011).

4.2.1.7 Scenario development

Historical trend analysis of the land-cover changes in the Jedeb shows an increasing demand for plantation forest, due probably to its use as a major source of firewood, lumber, house construction (both for people and for livestock) and various farm tools. This is especially true due to dwindling availability of the natural forests cover in the catchment. Recent regional and local policies prohibit the cutting of trees from natural forests, although this did not seem to have curbed deforestation. During field interviews it was learnt that a series of subsequent years of low yield motivates farmers to prefer planting trees such as Eucalyptus, which grow relatively fast and become a substitute cash earner. These plantation forests are often planted on degraded lands/steep slope area, and usually on higher elevation spots such as the hills. Natural woody vegetation exists almost exclusively on the riparian zones of the rivers and streams in the catchment as these are often unreachable and also unusable for other land uses due to deep river gorges and stony soils. Reduction in grassland impacts in particular the farmers with livestock. With growing population and livestock, peripheral grassland areas that were often left unused due to the unfriendly terrain are increasingly being used for cultivation and grazing, thereby exacerbating land degradation and soil erosion. It seems that, at least for the foreseeable future, this trend may not change much, including in terms of land-use policy and/or demands for the various land-uses. Thus, a business-as-usual scenario for population and livestock growth (and their associated demands for cultivation, settlement and grass/grazing land) was used for simulating the land-use model until 2025. This scenario assumes population and livestock growth rates to continue with the historical growth rates of 2.5 % and 1.5% per year, respectively (CSA, 2007; FAO, 2004b). The choice of 2025 is in line with the country's long term Growth and Transformation Plans (GTPs) which aims at the nation achieving a middle income status by 2025 (FDRE, 2011).

4.3   Results and Discussions

Land-use change and drivers

Analysis of changes in land-cover between 1986 and 2009 is shown in Table 4-3. As shown in this table, cultivated area and plantation forest increased from 54.4% and 0.3% in 1986 to 69.5% and 3.4%, respectively, in 2009 (see also Figure 4-3). On the other hand, natural woody vegetation and grassland decreased from 14.6% and 24.4% to 11.6% and 21.2%, respectively, in 2009.

Table 4-3. Land-use conversion matrix (1986-2009): conversion between land-use classes in km2 and percentage of total area (in brackets).

2009 \ 1986	Natural Woody Vegetation	Plantation Forest	Cultivated Land	Grassland	Others	Total 2009 [km² (%)]
Natural Woody Vegetation	6.59 (2.22)	0 (0)	2.38 (0.8)	2.05 (0.69)	0.56 (0.19)	11.58 (3.9)
Plantation Forest	1.40 (0.47)	0.89 (0.3)	3.77 (1.27)	3.86 (1.3)	0.18 (0.06)	10.1 (3.4)
Cultivated land	9.50 (3.2)	0 (0)	150.58 (50.7)	45.07 (15.18)	1.27 (0.42)	206.42 (69.5)
Grassland	25.84 (8.7)	0 (0)	4.51 (1.52)	20.87 (7.02)	11.74 (4)	62.96 (21.2)
Others	0.15 (0.05)	0 (0)	0.33 (0.11)	0.68 (0.23)	4.78 (1.6)	5.94 (2)
Total 1986 [km² (%)]	43.48 (14.64)	0.89 (0.3)	161.57 (54.4)	72.53 (24.42)	18.53 (6.2)	297 (100)

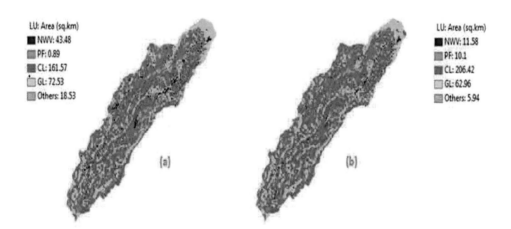

Figure 4-3. Observed land-use maps of (a) 1986 and (b) 2009.

From the analysis results shown in Table 4-3 and Figure 4-2, major changes were observed between 1986 and 2009, resulting, mainly, in an increase in cultivated land, and reduction in grassland and natural woody vegetation. Figure 4-4 shows map differences/changes in land-cover maps of 1986 and 2009.

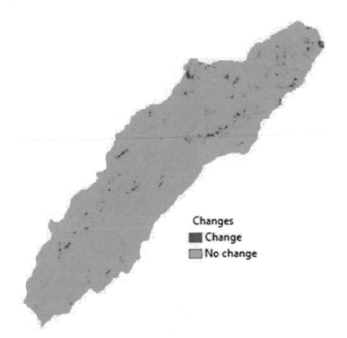

Figure 4-4. Changes between (a) observed maps of 2009 and 1986

The two major land-use change conversions were natural woody vegetation to grassland (close to 60% of the original woody vegetation is converted to grassland) and grassland to cultivated land (almost 60% of the original grassland is converted to cultivated land). On the other hand, land-use classes such as plantation forest did not seem to convert to other land-use types during the observed timeframe. Instead it seems that the plantation land-use type continued to expand as farmers increasingly change portions of their plots to plantation forest for fuelwood and construction materials, owing to the declining availability of and restrictive local policies on natural forest resources.

Table 4-4 presents summary of correlations between determinant variables land cover types. Looking at this table, we can see that the land-cover class 'Natural woody vegetation' is positively correlated with distances from forest edge and settlement whereas it is negatively correlated with slope and population. Cultivated land correlates strongly with population, slope, distance to market, distance to settlement. Grassland correlates with elevation, slope, distance to settlement, population and distance to water. Plantation forest shows strong correlation with slope, distance to road, distance to settlement, elevation and population. Similarly, the 'others' land use type (which includes urban, bare land and wetlands) has a slight correlations with population, distance to water and slope.

Table 4-4. Summary of significant correlations between land use and driving forces

LU / Variables	Pop.	Slope	Elev.	Dist. to settlm.	Dist.to Roads	Dist.to Market	Livestock	Dist.to Water	Dist.to forest edge
Natural woody vegetation	-0.23	-0.6	0.03	0.08	0.04	-0.01	0.02	0.02	0.84
Plantation forest	0.14	0.69	0.39	-0.52	-0.28	0.02	0.04	0.04	0.01
Cultivated land	0.79	0.65	0.02	-0.26	-0.04	-0.52	0.04	-0.05	0.001
Grassland	-0.4	0.68	0.78	-0.54	-0.03	-0.01	0.23	-0.31	0.04
Others	0.1	0.06	0.001	-0.02	0.01	0.03	0.01	0.1	0.002

The PCA method was conducted on the identified eleven potential land-use change drivers to determine the major explanatory variables of the change. Five components with eigenvalues >1 according to Kaiser's criterion (Kaiser, 1960) were retained. The rotated component loadings and communality estimates are shown in Table 4-5. The amount of variance in each driver variable that can be explained by the retained five components is represented by the communality estimates. From Table 4-5, we can see that component 1 (PC1) strongly correlates with population, distance to market, slope, and distance to settlement, explains 29.9% of the variance with high loadings (>0.7). PC2, which correlates with elevation and livestock, explained about 17% of the variance. Distance to road is correlated with PC3, which explains about 16.5% and distance to forest edge is strongly correlated with PC4, which explains about 11.3% of the variance. PC5, which strongly correlates to distance to water, explains about 10.8% of the variance. In combination, the five components explained about 85% of the change in land-use (Table 4-5).

Table 4-5. Factor loadings after varimax rotation and communality estimates (loadings >0.7 are in bold)

LU-drivers	Rotated component loadings					Communality
	PC1	PC2	PC3	PC4	PC5	estimates
Population	**0.887**	-0.227	0.418	-0.147	0.281	0.946
Distance to market	**-0.718**	0.005	0.145	-0.020	0.127	0.864
Distance to road	-0.135	0.001	**0.889**	0.308	0.054	0.885
Slope	**0.741**	-0.252	0.161	-0.312	0.319	0.939
Elevation	0.010	**0.929**	0.458	0.208	0.121	0.932
Livestock	0.320	**0.721**	0.120	0.089	0.073	0.786
				-		
Distance to settlement	**-0.753**	-0.549	-0.644	0.078	0.057	0.906
Distance to water	-0.247	0.114	0.324	0.096	**0.895**	0.917
Soil type	0.260	-0.022	0.081	0.069	0.151	0.671
Precipitation	0.253	0.001	0.059	0.066	0.173	0.681
Distance to forest edge	0.078	0.002	0.013	**0.921**	0.091	0.884
Initial eigenvalues	3.29	1.88	1.81	1.24	1.19	-
Variance (%)	29.91	17.09	16.45	11.27	10.82	-
Cumulative						-
variance(%)	29.91	47.00	63.45	74.73	85.55	

Land-use change rules

Comparing with summary of the initial correlations in Table 4-4, we see that components PC1 to PC5 in Table 4-5 are correlated with cultivated land, grassland, plantation forest, natural woody vegetation and 'others' land-use types, respectively. The results from the PCA loadings and the correlation table provide the basis for estimation of parameters of the initial suitability in SITE per the land-use types as shown in Table 4-6.

Initial weights computed and assigned for each determinant using Eq. 3.2 are shown on the 'Initial weight' column of Table 4-6, and its values are subject for calibration after which calibrated weights are replaced in the 'Assigned weight' column. The introduction of initial suitability weight values for the SITE modelling framework puts the model calibration into perspective with respect to field observations. Substantial divergence of calibrated values from the initial weights would mean that a recheck may be necessary. This way, chances for equifinality, a situation where a given state (level of model performance) can be reached by different potential combinations (variations of parameter sets), during model calibration can be avoided or at least minimized. On the other hand, the convergence or the closeness in value of the initial and assigned weights gives a certain level of confidence in the model parameterization and in the use of the resulting model for future scenario simulations.

Table 4-6. Land-use suitability rule-sets

Land use	Variable	Suitability ranges	Initial weight	Assigned weight (calibrated)	Direction of relationship *
Natural and Woody Vegetation	Distance to forest edge	>1,000m	0.35	0.5	Positive
	Distance to Roads	>5,000m	0.25	0.2	Positive
	Slope	<40%	0.2	0.2	Negative
	Distance to settlement	>3,000m	0.2	0.1	Positive
Cultivated land	Slope	<20%	0.4	0.66	-
	Distance to Settlement	<5,000m	0.1	0.2	Negative
	Distance to market	<10,000m	0.2	0.1	Negative
	Distance to water	<10,000m	0.3	0.24	Negative
Plantation Forest	Slope	5%-40%	0.2	0.3	-
	Elevation	1,200m-3,400m	0.2	0.1	-
	Distance to settlement	<5,000m	0.4	0.5	Negative
	Distance to road	<1,000m	0.2	0.1	Positive
Grassland	Slope	>10%	0.3	0.3	Positive
	Elevation	>2,600m.a.s.l	0.25	0.2	Positive
	Distance to water	<5,000m	0.2.5	0.3	Negative
	Distance to settlement	<20,000m	0.2	0.2	Negative

* Negative relationship type shows that as the value of the variable increase, the suitability of the variable for the land-use type will decrease and vice-versa. Positive relationship shows that as the value of the variable increases, the suitability increases as well. This relationship is an interpretation of the correlation analysis result presented in Tables 3-4 and 3-5.

Model evaluation

The land-use model, simulated from 1986 to 2009 using the assumed demands, land-use change drivers and the defined rule-sets, was calibrated and evaluated using indices of quantity and allocation disagreement measures. Quantity and allocation disagreement between the simulated and the observed cover maps of 2009 show an 8.7% quantity disagreement and a 7.3% allocation disagreement, adding up to a total disagreement of 16% between the two land cover maps (Table 4-7).

Table 4-7. Map comparison indices for the simulated and observed land-use of 2009

Name of Algorithm	Component	Measure (%)
Quantity & Allocation Disagreement	Change simulated as 'persistence' (quantity disagreement)	2.5
	Persistence simulated as 'change' (quantity disagreement)	6.2
	Change simulated as 'change to wrong category' (allocation disagreement)	7.3
	Total disagreement	16.0

From results of the model evaluation shown in Table 4-7, it was concluded that the simulated land-use map was able to mimic the land-use trends both in terms of allocation (spatial) as well as in terms of quantity. Although interpretations on levels of goodness of map comparisons remain still relatively subjective, the evaluation results showed an 84% agreement (more than the 80% threshold discussed previously corresponding to a 'very good' agreement). The developed land-use change model was, therefore, simulated up to 2025 for the scenario discussed earlier in this chapter. Figure 4-5 shows results of simulation between 2009 and 2025 and Figure 4-6 shows difference maps of the 2009 and 2025 land cover maps.

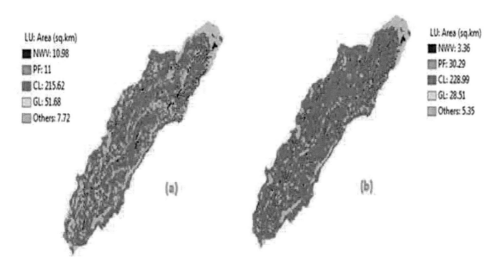

Figure 4-5. Simulated land-use maps of (a) 2009 and (b) 2025

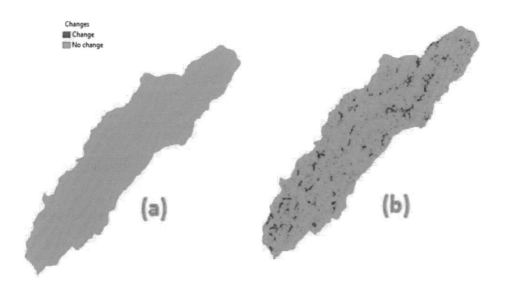

Figure 4-6. Difference map of (a) observed and simulated maps of 2009 and (b) simulated maps of 2009 and 2025

Table 4-8. Land-use conversion matrix (2009-2025): total area of conversion between land-use classes in km2 and percentages (in brackets)

2025 \ 2009	Natural Woody Vegetation	Plantation Forest	Cultivated Land	Grassland	Others	Total 2025 [km² (%)]
Natural Woody Vegetation	2.9 (0.98)	0.1 (0.03)	0.23 (0.08)	0.1 (0.03)	0.1 (0.03)	3.4 (1.15)
Plantation Forest	4 (1.35)	8.6 (2.9)	9.39 (3.16)	8.02 (2.7)	0.31 (0.1)	30.3 (10.2)
Cultivated land	1.8 (0.6)	0.4 (0.13)	196 (66)	28.34 (9.54)	2.76 (0.93)	229 (77.1)
Grassland	2.1 (0.7)	0.6 (0.2)	0.6 (0.2)	24.9 (8.4)	0.33 (0.11)	28.5 (9.6)
Others	0.78 (0.26)	0.4 (0.13)	0.2 (0.07)	1.6 (0.5)	2.44 (0.82)	5.4 (1.82)
Total 2009 [km² (%)]	11.58 (3.9)	10.1 (3.4)	206.42 (69.5)	62.96 (21.2)	5.94 (2)	297 (100)

The simulation results based on BAU scenario show that the expansion of the cultivation land will take about 77% of the total land cover by 2025 (Figure 4-5; Table 4-8). Compared to the period from 1986 to 2009 (54 in 1986 and 70 in 2009, see Table 4-3), the growth rate declines, from 200 to 141 ha/year. This may reflect the exhaustion of further suitable land for cultivation based on the defined suitability criteria. It seems likely that plantation forest area will nearly triple by 2025 at the expense of grassland and cultivated land (Table 4-8). Coverage of natural woody vegetation and grasslands continue to decline.

The scenario simulation results, as shown in Table 4-8 and Figure 4-5, can provide valuable insights on potential implications of land-use management and policy both from a local as well as a regional perspective. First, continuing decline of natural woody vegetation and grassland covers implies exacerbation in land degradation and soil erosion in the catchment (Bewket and Abebe, 2013; Simane et al., 2013). This can have local consequences such as reduction of environmental and ecological services, thereby impacting crop yields from cultivated lands. Second, the topography of the watershed as the source locations of the Upper Blue Nile River is dominated by rugged and mountainous landscapes. As a result this watershed has been described previously as prone to soil erosion and gully formation(Tekleab et al., 2014a; Tekleab et al., 2014b), which has led to decline in soil fertility and even loss of plots by local farmers(Haregeweyn et al., 2016; Zeleke and Hurni, 2001). The continuing decline of grasslands and natural vegetation, combined with expanding cultivation, would imply that the local erosion and gully formation phenomenon is bound to deteriorate unless more effective policies and management interventions are developed and implemented. At a more regional scale, consequences of the increase in erosion would imply an accumulation of more sediment in the downstream reservoirs. Simulation of land-use change scenario analysis such as in this study may inspire local as well as regional policy makers towards a coordinated regional land and water resources policy and management outlook.

Overall, the chapter showed that, in spite of the complexities of involving a wide range of socio-economic and biophysical factors in land-use modelling, the major trends of the past can be captured and reproduced to predict a likely trajectory of land-use change in the Jedeb catchment. As the land-use modelling presented in this chapter involves various socio-environmental parameters and complexities, a number of uncertainties will likely affect our model results. Besides uncertainties pertinent to the land-use model itself, those that propagate with the data gathered and used for the modelling can be expected to affect the certainty of results. We believe that the allocation of initial weight parameters (which then are checked against the assigned weights through calibration afterwards) help reduce such uncertainties. As to uncertainties involving data, we have tried to quantify as many of the variables as possible through various spatial correlation techniques with the hope of reducing subjectivity. Furthermore, where empirical data were lacking, assumptions were made based on local expert judgments and field observations.

4.4   Conclusion and Recommendations

Land-use change modelling involves complex layers of socio-economic and biophysical factors. With the objective of developing a predictive land-use model, we analyzed socio-economic and biophysical land-use drivers. We also developed a land-use change model that was parameterized and calibrated using field data. Based on a 1986 land cover as an initial map, we developed and simulated a land-use model until 2009. Then we evaluated and calibrated the simulated map of 2009 with a Landsat derived land cover map for the same year. The study demonstrates methods and techniques to identifying and analyzing land-use change drivers and field based parameterization of land-use change models. The simulated map of the year 2009, a result of a land-use change model with an initial land cover map of

1986, showed an overall good performance in mimicking trends and magnitudes of the observed land cover map. Once evaluated, the simulated model was further continued to simulate to the year 2025 under a business-as-usual scenario. This scenario assumes present rates of growth in population and livestock as well as associated demands to continue the same. The fact that no explicit water availability or constraint (except for distances from water bodies) was considered as a land-use change factor may be one of the limitations of this study. This may especially be true for a catchment like that of the Jedeb where river water is inaccessible (flows in deep gorges) and thus other hydrologic components such as surface runoff, ground water storage, and/or evapotranspiration may be better explanatory variables instead of distance to water sources. We believe that accounting for hydrologic impacts on the land-use dynamics of this catchment might improve understanding of the catchment land-use dynamics. A major recommendation from this chapter is, therefore, exploration of dynamic and semi-dynamic feedback between land-use change and hydrology which is explored in more detail in chapters 5 and 6.

PART II: FEEDBACK BETWEEN LAND USE AND HYDROLOGY

Land and water are not really separate things, but they are separate words.
- David Rains Wallace,
The Untamed Garden and Other Personal Essays

Chapter 5. Modelling hydrologic impacts of semi-dynamic land use in the Jedeb[8]

5.1   Introduction

Land use and hydrologic processes are believed to be interlinked whereby changes in land use affects hydrologic processes such as interception, evapotranspiration, infiltration, stream flow and runoff (Costa et al., 2003; Niehoff et al., 2002; Tong and Chen, 2002; Warburton et al., 2012). Evaluating and understanding effects of land use and land cover (LULC) changes on hydrologic responses of catchments is important particularly in the face of increasing population and associated environmental demands (Rientjes et al., 2011).

Changes in LULC involve complex socio-economic and biophysical processes: drivers and rates of changes in LULC are different from location to location, and from society to society. The use of static LULC map in hydrologic models as an input to simulating hydrologic responses ignores the fact that LULC is essentially dynamic. Catchment hydrology is, therefore, affected by direct or indirect changes in LULC and associated anthropogenic effects. Direct and indirect effects of natural and human-induced changes in LULC that can affect hydrology include river morphology (roughness), leaf area index, surface resistance, runoff curve number (CN), and rooting depth; all of which are important parameters in hydrologic modelling (Tang, 2016). Recent developments in interdisciplinary socio-environmental study related to land and water, known as 'socio-hydrology' (Elshafei et al., 2015; Gober and Wheater, 2015; Sivapalan et al., 2014), has highlighted the importance of anthropogenic effects on hydrology via proxies such as land-use. Socio-hydrology brings an interest in human values, markets, social organizations and public policy to the traditional emphasis of water science on climate and hydrology. As much as we believe this is an interesting development, the process-response representation of LULC changes in many hydrologic models employed for such analysis is still simplistic. LULC is taken either as a static input or as one that may be, depending on the design of the particular hydrologic model used for the analysis, set to change statistically. Thus, hydrologic responses to dynamic

---

[8] This chapter is based on material in: Yalew, S.G., van Griensven, A., van der Zaag, P, (2018). Hydrologic impacts of semi-dynamic land-use change in the Blue Nile basin. *Environment Systems and Decisions* (under review).

LULC are not well investigated in general, and in the Abbay (Upper Blue Nile) basin in particular.

In this chapter, we investigate hydrologic response to land-use change in the Jedeb catchment of the Abbay basin. Based on the level of dynamics (frequency) and direction of feedback

(whether one-way or two-way feedback) between land-use change and hydrologic models, we distinguished between 'semi-dynamic' and 'dynamic' land-use changes. The term 'semi-dynamic' feedback is used to denote that yearly land-use change simulated from the land-use model is provided as input to the hydrological model, but no feedback is received from the hydrologic model to influence the next round of the land-use model output. So this is essentially a one-way feedback. 'Dynamic' feedback is used to denote that both the land-use change and the hydrologic models exchange data at a yearly time-step and that thus, it is a two-way feedback. In this chapter, we investigated the semi-dynamic feedback between land-use change and hydrology. We use a physically based distributed hydrologic model and a dynamic land-use change model for modelling the hydrology and for producing yearly land-use maps, respectively.

5.2   Existing literature: state-of-the-art

The interaction between LULC and hydrologic responses of catchments may seem well investigated, given that a large number of articles are available in this topic: a simple 'Web of Science' search (Reuters, 2012) brings well over 500 articles using keywords " 'hydrologic response' AND 'land-use change' " alone. Articles covering hydrologic response to LULC change generally use either: i) static land-use at the beginning and end year of simulation, hence, using two land-use maps, one as a base and another as a reference map (Li et al., 2017; Li et al., 2007; Serpa et al., 2015; Yan et al., 2013), ii) three or more episodes of land-use map, such as decadal land-use maps or land-use maps from the beginning, middle and end year of simulation period (Bhaduri et al., 2000; Dang and Kawasaki, 2017; Nie et al., 2011; Tekleab et al., 2014a), iii) use a static land-use that can be changed statistically in terms of percentage at run time (Ghaffari et al., 2010; Mango et al., 2011; Sajikumar and Remya, 2015), or iv) dynamic land-use maps (Wagner et al., 2016). The last category, which is closest to the interest of our study, is rarely investigated. In fact, the only literature we could find at the time of this writing was Wagner et al. (2016) which integrated an urban land-use change model, SLEUTH (Clarke, 2008), and a hydrologic model, SWAT (Arnold et al.,

1998), to project water-yield in the city of Pune, India. However, even this was based on pre-planned/prepared land-use maps of yearly urban expansion scenarios, and thus it may barely quality as representative of the dynamic of yearly land-use changes resulting from land-use change model simulations based on interactions of various socio-environmental inputs and driving factors. This chapter tries to fill this gap by investigating effects of semi-dynamic simulations of land-use changes on responses of hydrologic components.

A number of articles have analyzed the effects of change of LULC on hydrologic responses such as streamflow and runoff generations in various catchments in the Abbay basin (Bewket and Sterk, 2005; Hurni et al., 2005; Koch et al., 2012; Rientjes et al., 2011; Tekleab et al., 2014a; Woldegiorgis, 2013). Bewket and Sterk (2005), using two episodes of land-use maps, reported a decrease of the dry season flows in Chemoga catchment (adjacent to the Jedeb catchment) which they attributed to changes in LULC. Rientjes et al. (2011), using two episodes of land-use maps, showed that both LULC and seasonal distribution of rainfall were the causes of changes in streamflow of the Gilgel Abbay River (the source region of the Abbay river). Using decadal land-use maps, Tekleab et al. (2014a) analyzed impacts of LULC change on various components of the hydrology: i) through evaluation of daily flow variability parameters to detect statistical significance of the change of the hydrologic response, and ii) through a conceptual monthly hydrologic model to detect changes in the model parameters over different periods (decadal changes from 1973 to 2010). Results showed changes in model parameters such as soil moisture over different periods, which, the authors explain, could be attributed to changes in LULC. Thus, Tekleab et al (2014a) concluded that there was a need for further research on impacts of changes LULC on streamflow in the Jedeb catchment using a (semi-)distributed physically based model that can account for the spatiotemporal variation of climatic and vegetation patterns in the catchment.

Koch et al (2012) and Woldegiorgis (2013) analyzed the hydrologic response of LULC changes in the Jedeb and Rib-Gumara catchments in the Abbay basin, respectively, using a script developed for this purpose (Koch et al., 2012). The script statistically interpolates in between land-use maps using land-use from two given periods. According to these authors, five land-use maps from 1957, 1974, 1986, 1994, and 2009 derived from aerial photographs and Landsat TM imageries were gathered and pre-processed for input to the Soil and Water Assessment Tool (SWAT) (Arnold et al., 1998). A python script was then developed that statistically generates in-between land-use maps which the authors describe are results of linear functions between two given model setups. Using the five aerial/satellite derived land-use maps, several in between (interpolated) land-use maps were produced and automatically fed to the SWAT model to see hydrologic impacts of the changes in the land use. The study reported that the representation of the changing (semi-dynamic, as per our earlier definition) land-use inputs in SWAT resulted in higher runoff especially during the peak flow season compared to that of the normal model setup. However, although this approach produces yearly land-use inputs to the hydrologic model, the land-use inputs are entirely based on simple interpolation between maps. It accounts for neither socio-economic nor biophysical changes that can influence land-use changes.

A more recent study on the analysis of effects of land-use change on hydrologic response was conducted by Woldesenbet et al. (2017) in two watersheds in the Abbay basin using two episodes of land-use maps. The authors concluded that expansion of agricultural land replacing natural vegetation contributed to high run-off in the study watersheds. Other studies on the impacts of (static) land-use on hydrologic responses on various catchments reported different, and sometimes contradictory results (Costa et al., 2003; Rientjes et al., 2011; Siriwardena et al., 2006). Assessing effects of (semi-)dynamic land-use changes which are normally derived by environmental and anthropogenic factors on hydrologic responses remains an important subject that is least investigated in general and in the Abbay basin in particular.

In this study, we present a new approach for the analysis of effects of land-use change on hydrologic components. Instead of taking a static land-use map or few episodes of land-use maps or using varying land-use proportions in hydrologic models, a separate land-use change model based on various socio-environmental land-use change drivers was developed. Yearly maps of this land-use change model were used to update the land-use input in the hydrologic model, thereby incorporating anthropogenic effects on hydrologic response. The following sections describe the study area and the method used.

5.3   Materials and methods

Study area

The Jedeb catchment is situated in the south-west part of Mount Choke and it is part of the headwater of the Abbay (Upper Blue Nile) basin (Figure 4.1). It covers an area of 297 km^2 and is situated between 10°22′ to 10°40′ N and 37°33′ to 37°50′ E. The area is known for its diverse topography with elevation extending from 2100 to 4000 m.a.s.l., and slopes ranging from nearly flat to very steep (>45°). The mean annual rainfall varies between 1400 and 1600 mm per annum (based on data from three climate stations: Debre Markos, Anjeni and Rob Gebeya). The steep slopes, coupled with erosive rains, have contributed to excessively high rates of land degradation and soil erosion (Betrie et al., 2011; Easton et al., 2010). As one of the severely eroded and degraded parts of the basin, the catchment received the attention of researchers who undertake various socio-environmental and water resources studies in the catchment (Teferi et al., 2013a; Tesfaye and Brouwer, 2012). Land-use and land-cover changes, such as loss of grassland cover due to overgrazing, poor land-use management and conversion of grassland to agricultural land for instance, may have contributed to a higher level of gully formation, soil erosion and land degradation in general. Between 1957 and 2009, 46% of the watershed has undergone land-use changes without proper soil and water conservation measures in place (Bewket and Teferi, 2009). The changes in land use and land cover are thought to be among the major causes of high erosion rates in the basin (Bewket and Teferi, 2009; Pankhurst, 2010). Whether this trend will continue is dependent, among other things, on future land use.

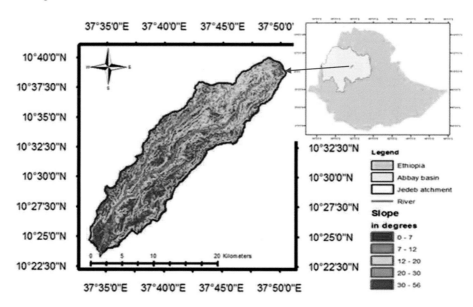

Figure 5-1. Location and topographic map of the Jedeb catchment in the Abbay (Upper Blue Nile) basin, Ethiopia

Hydrological model

For modelling the hydrology, we used the WFlow framework (Schellekens, 2014). Wflow is a physically based distributed hydrologic modelling framework that maximizes the use of available spatial data (Schellekens, 2014). The model, derived from the CQFlow model (Köhler et al., 2006), is built on a python based geographical information systems (GIS) environment called PCRaster (Van Deursen, 1995). The structure of the model is transparent and can be changed by other modelers easily. It also allows for a rapid application development of 'new' models (Schellekens, 2014). Figure 4-2 shows an overview of the different processes and fluxes represented in WFlow.

Some of the key features of this framework are:

i.   Interception is modelled using the Gash model (Gash, 1979) for rainfall interception that works with daily time steps.

ii. The model uses potential evapotranspiration as input time series and derives the actual evaporation based on soil water content and vegetation type.

iii. The soil is represented using a simple bucket model that assumes an exponential decay of the saturated conductivity (Ksat) with depth, a soil processes schematization based on the TOPOG_SBM model (Vertessy and Elsenbeer, 1999).

iv. Surface runoff is modelled using a kinematic wave routine.

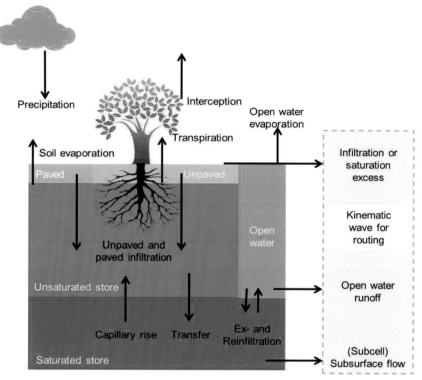

Figure 5-2. Overview of the different processes and fluxes in the WFlow model
(Schellekens, 2014)

WFlow is organized as directories and sub-directories. A 'case' is a directory holding all the
data and initial parameters needed to run the model. WFlow input data are divided into two:
static and dynamic. Static inputs, such as DEM, soil and land use (if static), are placed in a
'staticmaps' directory. Dynamic inputs, such as precipitation and evapotranspiration data are
placed in an 'inmaps' folder. Parameter and flow data are placed in 'intbl' and 'intss' folders,
respectively. Within a case the model output (the results) are stored in a separate directory,
'Run'.

Data and data processing

The actual data requirements of the modelling framework depend on the application of the
model. The model inputs used in this study are listed below.

• Static data:

   • Digital Elevation Model (DEM)

   • A land-use map (the one given as static input)

- A map representing soil physical parameters/soil map.
- Dynamic data (spatial time series, map-stacks)
  - Precipitation
  - Potential evapotranspiration
  - Temperature
  - Land-use maps (semi-dynamic, i.e., map-stacks for the dynamic simulation)
- Model parameters (per land use and soil type combination)
  - Rooting depth, Manning's N, canopy storage, soil infiltration capacity, etc.

5.3.1.1 Land cover

We used a modified version of the land-use model developed in Chapter 4 (Yalew et al., 2016b) for producing yearly land-use change maps to be used as input the hydrologic model. For the development of the land-use model, first an initial land-use map of 1986 derived from Landsat TM images, and described in an earlier study (Teferi et al., 2013a), was used as a base map for the simulation. Then the model was simulated until 2012. Two reference land-use maps from 1989 and 2009, derived from Landsat TM images similar to the 1986, were used as reference maps for comparison against the simulated maps for the respective years (Figure 4-3). These Landsat maps were thoroughly processed and validated during a previous study of the catchment's land cover (Teferi et al., 2013a). Once the land-use model was calibrated and evaluated using the reference maps, the simulation maps of this model from 1989-2012 are used as yearly semi-dynamic inputs to the WFlow hydrologic model.

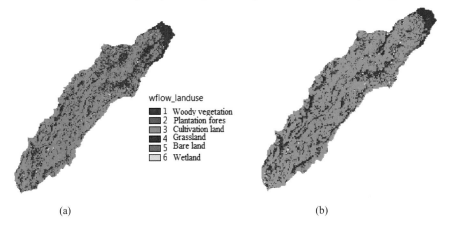

wflow_landuse
■ 1 Woody vegetation
■ 2 Plantation fores
□ 3 Cultivation land
■ 4 Grassland
■ 5 Bare land
□ 6 Wetland

(a)   (b)

Figure 5-3. Simulated land-use maps of 1989 (a) and 2012 (b).

5.3.1.2 Soil

Soil map for the catchment was taken from the global Harmonized World Soil Database (HWSD v1.2) (Fischer et al., 2008) which has a resolution of 3 arc seconds (approximately 1km) and the Digital Soil Map of the World (DSMW) (FAO, 2007) with a resolution of 5 arc-minutes (approximately 10km). Additional soil parameters, including soil rooting depth, were taken from NASA's ISLSCP II dataset (Schenk and Jackson, 2009), see Figure 4-4.

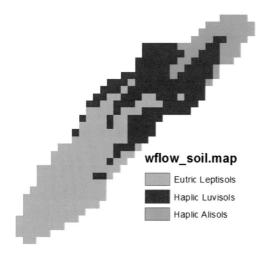

Figure 5-4. Soil groups in the Jedeb catchment

5.3.1.3 Weather data

Streamflow data was collected from the Ministry of Water Resources and Energy of Ethiopia, whereas daily precipitation and temperature data was obtained from the National Meteorological Agency of Ethiopia (Table 4-1).

Table 5-1. Record of daily hydro-meteorological data in the Jedeb catchment

Variable	Station	Record period	Long term annual mean	Coef. variation (%)	Missing data (%)
Precip.	Debre Markos	1954-2012	1326 (mm a-1)	12	<1
Precip.	Rob Gebeya	1989-2012	1434 (mm a-1)	14	0
Temp.	Debre Markos	1973-2012	16.3 (°C)	3.5	2
Discharge	Jedeb	1973-2012	593 (mm a-1)	24	5

Parameter data

Input parameters required for the WFlow interception module (e.g., maximum canopy storage) and for the soil module (e.g., infiltration capacity and Manning's N) were estimated per soil class based on literature and global soil parameter databases (see Appendix 1).

Catchment delineation

Two python scripts are provided with the WFflow modelling framework for catchment delineation: 'wflow_prepare_step1.py' and 'wflow_prepare_step2.py'. The scripts do the catchment delineation through terrain analyses (Figure 4-5). They calculate slopes from DEM and compute flow direction and flow accumulation taking into account the geographical location of gauges and river networks. Once catchment delineation is performed, the WFlow model is run from a command line interface.

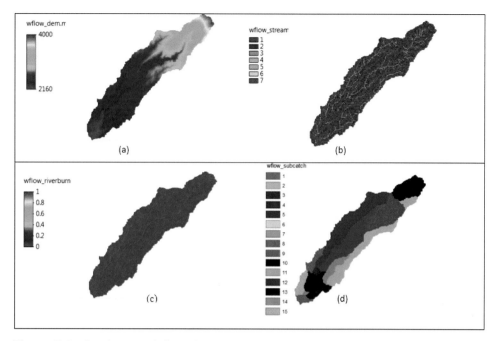

Figure 5-5. Catchment delineation in WFlow: (a) digital elevation map (DEM) (b) stream order as a result of computation of flow accumulation (c) burn-in of river network (d) resulting sub-catchments

Calibration and validation

The land-use model was calibrated with the GALib genetic algorithm library (Wall, 1996) which comes integrated within SITE. CALib is used for the auto-calibration of the land-use model based on defined suitability, demand for various land-uses and allocation routines. The land-use simulation was evaluated ring the simulated land-use maps of 1989 and 2012 against reference land-use maps for the same years using the quantity disagreement (QD) and allocation disagreement (AD) measures (Pontius and Millones, 2011). These measures summarize cross-tabulation matrix of two categorical maps (simulated and reference) in terms of quantity and spatial allocation disagreements and are reported to sufficiently account for differences in terms of the quantity (changes or persistence) and allocation of matching land-use pixels (Pontius and Millones, 2011). A threshold for a generally acceptable level of accuracy/agreement in map comparisons does not exist in the literature. Although not particularly for QD and AD, Landis and Koch (1977a) suggested agreement ranges of map comparison results into three groups: agreement value greater than 0.8 (80%) represents strong agreement; agreement value between 0.4 (40%) and 0.8 (80%) represent moderate agreement; and agreement value less than 0.4 (40%) represents poor agreement between two maps. We used these guides to evaluate the performance comparison between the simulated and the reference land-use maps.

With regards to the hydrologic model, the main parameters for calibration in WFlow include *M*-soil parameters determining the decrease of saturated conductivity with depth taken from a global dataset (Fischer et al., 2008), *N*-Manning's roughness parameter taken from literature (Koch et al., 2012), *Ksat*-saturated conductivity of the store at the surface taken from literature (see Appendix 1.2), and *FirstZoneCapacity*-maximum capacity of the saturated store taken from literature (see Appendix 1.2). The WFlow calibration and validation performance was evaluated using the Nash–Sutcliffe efficiency (NSE) (Nash and Sutcliffe, 1970) coefficient and the PBIAS (Moriasi et al., 2007) measures. NSE and PBIAS measures are presented below using equations 6-1 and 6-2, respectively.

$$NS = 1 - \frac{\sum_{t=1}^{T}(Q_{obs}^{t} - Q_{sim}^{t})^{2}}{\sum_{t=1}^{T}(Q_{obs}^{t} - Q_{mean}^{t})^{2}} \qquad (6.1)$$

where Q_{obs} is observed discharges; Q_{sim} is modeled discharge; Q_{mean} is mean observed discharge at time *t*.

PBIAS is an error index describing the average tendency of simulated values to be larger or smaller than observed data (Moriasi et al., 2007).

$$PBIAS = \frac{\sum_{j}(Sim_j - Obs_j)}{\sum_{j}(Obs_j)} * 100\% \qquad (6.2)$$

Model performances are generally considered satisfactory if NSE>0.5 and PBIAS< ±25% (Moriasi et al., 2007).

The WFlow model was calibrated for the years 1989-1993 and validated for the years1994-1998 using a static land-use input. The model performance for calibration and validation was evaluated using NSE and PBIAS measures. Then, the performance of the hydrologic model was evaluated for the years 1999-2012 using both static and semi-dynamic land-use inputs.

5.4   Results and Discussion

AD and QD measures between the simulated and reference land-use maps of 1989 show 8.2% and 7.1%, respectively, whereas 7.3% and 8.7% for AD and QD measures, respectively, were observed between the simulated and the reference land-use maps of 2009 (Table 4-2). Thus, the comparison shows a total disagreement of 15.3% (or a total agreement of 84.7%) between the simulated and reference land-use maps of 2009 and a total disagreement of 16% (or a total agreement of 84%) between the simulated and reference land-use maps of 2009.

Table 5-2 Comparison of simulated and reference land-use maps for the years 1989 and 2009

Land-use	Quantity and allocation disagreements			
	Components			Measure (%)
	Change simulated as 'persistence' (QD)	Persistence simulated as 'change' (QD)	Change simulated as 'change to wrong category' (AD)	Total disagreement
1989	2.7	4.4	8.2	15.3
2009	2.5	6.2	7.3	16

NSE and PBIAS performance measures for the calibration period of the hydrological model show values of 0.72 and -0.19, whereas NSE and PBIAS values of 0.68 and -0.22, respectively, were observed for the evaluation period (see figures 4-6, 4-7 and Table 4-3).

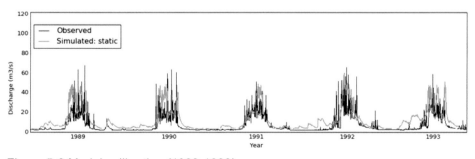

Figure 5-6 Model calibration (1989-1993)

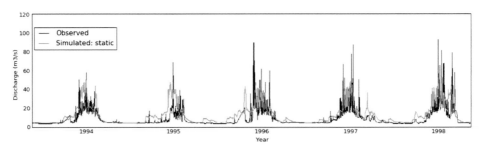

Figure 5-7. Model validation (1994-1998)

Table 5-3. WFlow model performance with static vs. semi-dynamic land-use inputs

Model	Land-use input	Calibration (1989-1993)		Validation (1994-1998)		Validation (1999-2012)	
		NSE	PBIAS	NSE	PBIAS	NSE	PBIAS
WFlow	Static	0.72	-0.19	0.68	-0.22	0.63	-0.20
	Semi-dynamic					0.7	-0.16

Once WFlow was calibrated and validated for the period 1989-1998, using a static land-use map of 1989, the model was evaluated using the same performance measures, NSE and PBIAS for the rest of the simulation period, i.e., from 1999-2012 both with static and semi-dynamic land-use inputs. The model with static land-use input shows NSE and PBIAS values of 0.63 and -0.20, respectively, whereas the model with semi-dynamic land-use inputs show NSE and PBIAS value of 0.7 and -0.16, respectively, for the period 1999-2012 (Table 4-3 and Figure 4-8). A closer look at the 'blue-box' part in Figure 4-8 showing zoomed-in comparison between observed and simulated flow hydrographs is presented in Figure 4-9.

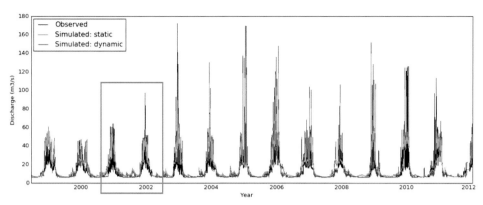

Figure 5-8. Observed vs. simulated discharge; the part in the 'blue-box' is further explored in Figure 9.

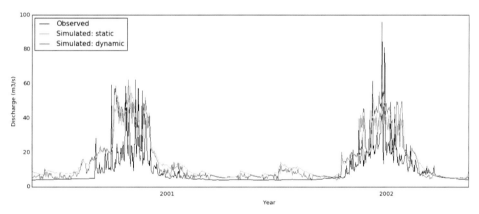

Figure 5-9. A closer look at observed and simulated flows with semi-dynamic and static land-use maps

Segmenting the flow into three regimes (high, average and low-flow), we presented overall average difference between the observed and the simulated flows using both static and semi-dynamic land-use inputs (Table 4-4).

Table 5-4. Overall average comparison of the simulated and observed flows (% difference) from 1999-2012

Flow regimes*	Land-use input for WFlow	
	static	semi-dynamic
High-flow	-19	-13
Average-flow	2.2	1.8
Low-flow	8.6	4.5

*N.B. High-flow is represented by flows of 40 m3/s and above, average-flow is represented by flows from 20 m3/s to 40 m3/s, and low-flow is represented by flows below 20 m3/s, all from observation of the streamflow hydrograph.

5.5   Discussion

The land-use change model was simulated from 1986-2012. Besides the base land-use map of 1986, two other reference land-use maps from 1989 and 2009 were used for comparison against the simulated land-use maps of the respective years. The comparison results (Table 4-2), using the quantity and allocation disagreement measures, show that the simulated land-use map of 1989 was in agreement with the reference map of the same year by about 84.7% (total disagreement of 15.3%) and the simulated map from 2009 was 84% (16% total disagreement) in agreement with that of the reference map of the same year. This comparison result represents a strong agreement according to Landis and Koch (1977a). Similarly, the WFlow hydrologic model was calibrated for 1989-1993 and validated for 1994-1998 using a static land-use map of 1989. NSE and PBIAS values, respectively, of 0.72 and -0.19 for the calibration period and 0.68 and -0.22 for the validation period were observed (Figures 4-6 and 4-6; Table 4-3). Both the NSE and PBIAS values were within suggested ranges for good model performance (Moriasi et al., 2007).

Once both the land-use and the hydrologic simulation models were calibrated and validated, yearly outputs of the land-use change simulation 1989-2012 were semi-dynamically used as inputs in the hydrologic model. Then, the hydrologic model was evaluated again for the period 1999-2012 using both static and semi-dynamic land-use inputs. NSE and PBIAS values of 0.63 and -0.20 were observed for the model with static land-use input whereas 0.7 and -0.16, respectively, where observed with the semi-dynamic land-use input. As is evident from Table 4-3, the hydrologic model with semi-dynamic land-use input shows a better model performance in both NSE and PBIAS measures for the period 1999-2012, also as can be observed from the hydrograph in Figure 4-8.

A closer look at the observed and simulated flow hydrographs (Figure 4-9) shows that the static/semi-dynamic land-use inputs have different effects on the various segments/regimes of the flow hydrograph. The first observation from the figure is that the simulated flow with both static and semi-dynamic land-use inputs is underestimated during high-flows and overestimated during low-flows, whereas it remained comparable during average-flows (Table 4-4). During the wet season, rainfall is at its peak, and a large amount of the stream flow is generated from surface runoff. The underlying assumption regarding the interaction between land-use and hydrology is that land with little vegetation cover is subject to high (and quick generation of) surface runoff, low infiltration rate and reduced groundwater recharge leading to higher volumes during peak-flows (Bewket and Sterk, 2005; Foley et al., 2005). In this perspective, the underestimation of the simulated flow suggests that the vegetation cover in the land-use maps used as input in to the hydrologic simulation (and where precipitation has to fulfil the evaporation, transpiration and soil moisture storage demands before generating runoff) might have been overestimated compared to the actual vegetation covers in the catchment, but that the semi-dynamically changing land-use maps reduce this error. Another way of putting this is that agricultural land, which increases surface runoff, might have expanded more than what was represented using either the static or the semi-dynamic land-use maps. The overestimation of the simulated flow compared to the observation during the dry (low-flow) seasons may partly be attributed to the increasing use of ground water abstraction to be expected from the increased human and livestock populations in the region (Abera et al., 2017; Ali et al., 2014). In addition, the natural vegetation cover in the catchment in particular and in the basin in general is being replaced with eucalyptus (Eucalyptus grandis and Pinus) trees (Gebrehiwot, 2015; Tekleab et al., 2014a). Various studies have reported that eucalyptus plantations can cause drastic reductions in stream flows (Bewket and Sterk, 2005; Scott and Lesch, 1997). Thus lower volume of the observed flow during low-flow seasons compared with the simulated flows can be partly attributed to the increased water consumption of the eucalyptus cover.

The second observation concerns the difference in flow response of the hydrologic model to static and semi-dynamic land uses. Although the general trend is that the simulated flows underestimate high-flows and overestimate low-flows, another clear trend is observed between the flow responses when using static and semi-dynamic land-use inputs (Figure 4-9; Table 4-4). The model with semi-dynamic land-use input shows better performance compared to the one with static input at mimicking the observation during the high and the low flow seasons. The aggregated difference between the simulated flow with static input and the observed flow during high and low-flow seasons, respectively, was -19% and 8.6%. The simulated flow using semi-dynamic land-use input showed -13% and 4.5% aggregated difference with the observed flow during high and low-flow seasons, respectively, whereas the two simulated flows were comparable on average-flow seasons. The overall percentage difference between the observed and the simulated flows was significantly smaller when the semi-dynamic land-use input was used. This is due to the fact that the gradually changing land use is better captured and thus simulated actual land-use mode accurately based on various biophysical and socio-economic changes including increasing population, and demands on land and water resources. From these results, it is evident that LULC change generally affects hydrologic response of catchments, and that the impact of semi-dynamic LULC change on hydrologic response when compared to static LULC is pronounced during high and low-flow periods. If hydrological models that assume LULC to remain static are used in dynamically changing landscapes, a systematic error is introduced.

5.6   Conclusion

Hydrologic models are often setup with a single/static land-use map to simulate long term simulations with only proportional/statistical changes in land-use. This, we argued in this chapter, leaves out the presentation of spatial socioeconomic and biophysical dynamics applicable to local conditions. A physically based distributed hydrologic model (WFlow) was setup for the period from 1989-2012 to test streamflow response to static and semi-dynamic land-use inputs in the Jedeb catchment model. The hydrologic model was simulated by taking semi-dynamic yearly land-use maps, a result of a separate land-use change model (SITE), for the period 1989-2012. Flow responses of the hydrologic model using both static and semi-dynamic land-use inputs were evaluated. The hydrologic model using semi-dynamic land-use input showed a better model performance and particularly at capturing the observed flow during the high-flow and low-flow periods. From these results, we conclude that LULC generally affects hydrologic response of catchments, and that the impact of semi-dynamic LULC change on the hydrology is evident from its pronounced effect on flow responses during high and low-flow periods. Thus, hydrological models of catchments that experience significant LULC change can benefit from dynamic land-use change.

Chapter 6. Feedback between coupled land-use and hydrologic models[9]

6.1 Introduction

In chapter 2, we discussed that land-use modelling involves various socio-environmental factors. Changes in land-use can also affect ecosystem services linked to LULC, thereby affecting sustainability of natural resources management. Sustainable natural resources management charts a path for economic development while sustaining environmental resources necessary to support it. The concept of ecosystem services provides a framework on which one can evaluate the challenges and opportunities of socio-environmental sustainability. Ecosystem services are defined as services people derive from their environment (Boyd and Banzhaf, 2007; Norgaard, 2010; Wallace, 2007). The concept integrates biophysical and socio-economic components and can be used to account for all benefits provided by the environment, enabling managers and decision-makers to understand the impacts of a development, policy, programmers or plans at different spatial and temporal scales (Johnston et al., 2014). It further enables managers to make informed decisions about trade-offs and synergies between various socio-environmental indicators for safeguarding livelihoods and promoting sustainability. Changes in land use and land cover (LULC) cause associated changes in ecosystem services provision and sustainability (Koschke et al., 2012; Lambin and Meyfroidt, 2010; Metzger et al., 2006; Reyers et al., 2009). For a grassland dominated region, for instance, changes in the grassland cover could change ecosystem services associated with grassland such as grazing, water retention/regulation, and erosion prevention. It has also been noted that poorly managed livestock grazing can lead to the emergence of regional syndromes inherent to global climate change: desertification, woody encroachment and deforestation (Asner et al., 2004; Mabbutt, 1984).

Land-use and hydrology are strongly interlinked whereby changes in land use affects hydrologic processes including interception, evapotranspiration, infiltration, stream flow and runoff (Costa et al., 2003; Niehoff et al., 2002; Tong and Chen, 2002; Warburton et al., 2012). Similarly, changes in hydrologic processes can influence the distribution and availability of water resources, which in turn can influence processes driving land-use and

---

[9] This chapter is based on: S.G.Yalew , T. Pilz, C. Schweitzer, S. Liersch, J. van der Kwast, A. van Griensven, M.L. Mul, P. van der Zaag. (2017) Coupling land-use change and hydrologic models for catchment ecosystem services. *Environmental Modeling & Software* (under review).

land cover changes. Changes in hydrologic components, for example, impact parameters associated with land-use suitability for agriculture including soil water balance, leaf area index, vegetation/crop growing seasons, vegetation root depth and root mass distributions (Calder, 1998; DeFries et al., 2004). Modelling the dynamic interactions between these sub-disciplines is an important endeavor which deserves more attention in the environmental modelling community (DeFries and Eshleman, 2004; Wagner et al., 2017). Most modelling practices on interactions of land-use and hydrology using respective models display notable limitations, however. Four general limitations are observed: (1) a static land-use map is used for an entire simulation period in most hydrologic models and modelling practices. Such representation of land-use is problematic because rates and magnitudes of land-use changes are dynamic in practice and the changes can be significant over a modelling period. Apart from biophysical factors such as weather and climate variability, land-use dynamics can also be driven by various social, economic and spatial factors including population, values of commodities, distances from various resources, infrastructures, and services (Lambin et al., 2003; Lambin et al., 2001; Verburg et al., 2004). (2) Hydrologic models are often developed to work only with biophysical inputs such as slope, soil, land use, and weather datasets. This leaves out aspects of socio-economics that contribute to changes in hydrologic processes and water resources directly or indirectly through, for instance, growing population and associated pressures on land-cover and land-use demands. If no socio-economic aspects are considered in hydrologic models, it would necessarily imply that any two catchments whose biophysical input components are equivalent will be assumed to have the same rates of changes and projections irrespective of specific environmental and socio-economic dynamics (e.g. population, density of settlement, in/out-migration, lifestyle, and demands for spatially relevant commodities such as land). (3) Likewise, most land-use change models ignore hydrologic components altogether in their simulations, and those which consider it, take limited proxies of hydrology/water resources using variables, such as average precipitation or distance from water sources, as primary inputs. This results in an oversimplification which ignores all other factors including rainfall intensities, infiltration rates, runoff components, and evapotranspiration, which can influence soil water balance, land suitability and productivity. Ignoring such hydrologic components and discounting water availability in land-use modelling can result in an unconstrained model with regards to potential or actual water resources availability and, thus, can lead to misleading conclusions. (4) On the other hand, even though modelers are aware of the limitations mentioned, incorporating all the depth and breadth of specialized disciplines of hydrology and land-use change into one model can be overly complex.

Integrated modelling methods have been advocated to address, at least partly, the problems mentioned above. Their holistic approach to socio-environmental problem solving in general and in land and water resources management has especially been an important attribute (Hamilton et al., 2015; Jakeman et al., 2013; Laniak et al., 2013). In line with this philosophy, an overarching discipline in the water domain, known as socio-hydrology (Sivapalan et al., 2012; Wheater and Gober, 2011), has emerged recently that emphasizes exploration of integrated socio-economic and anthropogenic feedbacks between land-use change and hydrology (Di Baldassarre et al., 2015).

To be sure, integrated modelling comes with its own challenges. One of the main challenges is associated with calibration and evaluation mechanisms in integrated subsystems. Changes in variables that used to impact only relatively some part of a subsystem can propagate throughout the whole integrated system (Voinov and Shugart, 2013). A more 'conservative' approach of integrated modelling is often adopted in many interdisciplinary modelling practices where specialized models from various disciplines are calibrated and evaluated independently and exchange data with other models through coupled systems. Model coupling combines specialized models in their entirety instead of relying on simplifications of the specialized models within an integrated framework. Proponents of this approach argue that the coupling approach enables a greater degree of transparency and accuracy in integrated models landscape and watershed models (Robinson et al., 2017; Verburg et al., 2016).

Landscapes provide a number of provisioning, supporting, regulating and cultural ecosystem services (Millennium Ecosystem Assessment, 2005). Agricultural intensification and extensification has enabled substantial increases in provisioning services (e.g. food production) by exploiting available land and water resources, but it has also transformed and degraded natural watersheds and landscapes. To counterbalance the negative effects of intensive agriculture, there is increasing interest in multifunctional landscapes and watersheds for sustainable use natural resources and associated supporting, regulating and cultural ecosystem services (Scherr and McNeely, 2008). Ecosystem services are often evaluated and quantified in association with land and water resources. Grasslands provide ecosystem services in the form of animal feeds/grazing, erosion control, water regulation, soil carbon retention, and biodiversity conservation (Lemaire et al., 2011; White et al., 2000). It is expected that, in the face of climate change and growing demands for agricultural land and productivity, future pressures on grassland ecosystems will intensify (Watkinson and Ormerod, 2001).

The main objective of this chapter is to evaluate whether dynamic feedback between land-use change and hydrologic models can improve performances of the respective models and/or whether it can produce a more realistic quantification of catchment ecosystem services. We coupled and tested the effect of dynamic feedback between two respective models: SITE and, SWIM (Soil and Water Integrated Model). The approach is applied to the Thukela catchment, South Africa, as a continuation of our prior experiment on the identification, valuation and mapping of various ecosystem services in the catchment (Yalew et al., 2014). Specifically, this study investigates the effect of model coupling on the sustainability of one of the common ecosystem services of grasslands in the catchment, namely grazing. Quantifying interactions of grassland ecosystem services and water resources in the catchment allows an evaluation of sustainable grazing levels. The evaluation is based on performance criteria for the coupled and uncoupled model results and on the importance of the coupling for the assessment of ecosystem services.

6.2 Study Area

As described in chapter 1, the Thukela catchment comprises of the Thukela district municipality in the KwaZulu Natal province, South Africa. The district comprises a marked biophysical gradient and diversity of habitat types that is determined by altitude, slope position, aspect, climate, topography and geology. High livestock population and poor soil and land management, among others, are causes attributable to degradation of the grasslands in the catchment. Increasing demand for arable and urban land decreases the extent of the grazing lands. Overgrazing, agricultural and urban expansion, mining, and poor land-use management practices are reported to have resulted in land and soil degradation as well as decrease in the quality and quantity of grasslands in the catchment. Grassland biome, which forms a large and important component of South African vegetation (Scott-Shaw and Schulze, 2013), continues to fragment, thereby impacting the most common grassland ecosystem service, grazing. On the other hand, livestock holds major economic and social values for the communal farmers in the catchment and thus grassland ecosystem service will continue to be essential. Through the National Environmental Management Act (RSA, 1998)

(NEMA, Act No. 107 of 1998), South Africa has acknowledged the need for sustainable development and socio-environmental management. As one of the grassland dominated districts in South Africa, the Thukela catchment presents various challenges related especially to the grazing ecosystem services of the grassland in the catchment. Identification, assessment and mapping of grassland biomass that can be used for sustainable grazing by livestock in this area are, therefore, critical for long term rangeland policy and agro-hydrological decision making in the catchment.

6.3   Materials and Methods

First, a hydrologic model is developed for a South African catchment and simulated for 30 years (1990-2010) using the Soil and Water Integrated Model (SWIM) (Krysanova et al., 1998). A separate land-use change model using SITE is likewise developed, calibrated and evaluated for the same catchment. Then, the two models are functionally integrated through coupling in a way that the output from the land-use change model is used to update inputs of the hydrologic model and vice-versa at a yearly time-step. Individual model inputs, model setup and structure as well as the method used for coupling the two models are discussed in the following subsections.

The Hydrological Model (SWIM)

We used the SWIM model (Krysanova et al., 1998) for hydrologic modelling of the catchment. SWIM is an open-source model used to simulate eco-hydrologic processes such as runoff generation, nutrient and carbon cycling, river discharge, plant growth, crop/biomass yields and erosion (Krysanova and Wechsung, 2000). It can simulate agricultural management, feedbacks of climate and land-use changes as well as dynamic vegetation growth processes (Krysanova et al., 1998). SWIM takes meteorological, topographic, soil and land-use datasets as inputs. It operates at a daily time-step and uses a three-level spatial discretization scheme: watershed, sub-basin and hydrotopes. Sub-basins are delineated from digital elevation model (DEM) and represent small individual watersheds whereas hydrotopes are small hydrologic units with similar land use, vegetation and soil types (Figure 6-1).

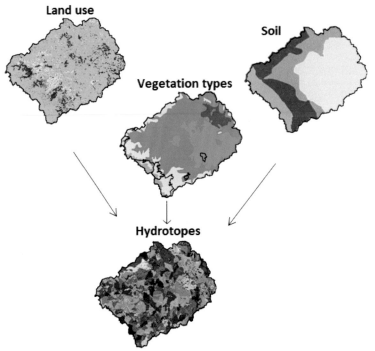

Figure 6-1. Setup of hydrotope units in SWIM

6.3.1.1 Inputs and model setup

Inputs for the SWIM model include soil, land use, stream flow, and meteorological data including precipitation, temperature, solar radiation, and relative humidity. Soil information was derived from the Harmonized World Soil Database (FAO et al., 2012). Meteorological information was taken from the Water and Global Change (WATCH) project (Weedon et al., 2010). Streamflow data was derived from the Global Runoff Data Centre (GRDC, 2013), and from the South African Department of Water Affairs (DWA, 2013). Hydrology in SWIM is governed by the water balance equation that is calculated at the hydrotope level, Eq. 6-1.

$$\Delta S_t = S_t - S_{t-1}$$

(6.1)

where ΔS_t is given by the difference in the amount of water stored in the soil between time t, and time $t-1$.

The biomass produced after each hydrologic year by SWIM is used as input by SITE. For that, a specific number of potential heat units required for maturity has to be defined for each vegetation/crop type in SWIM. The computation of biomass accumulation follows the approach of Monteith et al. (1977). Photosynthetic active radiation (P_{AR}) estimated from input solar radiation and leaf area index (LAI) is used to calculate potential biomass accumulation, Eq. 6-2.

$$\Delta B_P = B_E * P_{AR}$$

(6.2)

where BE is a crop specific factor converting energy to biomass. ΔBP is then reduced to get the daily actual biomass increase, Eq. 6-3.

$$\Delta B = \Delta B_P * R_{EGF}$$

(6.3)

REGF is a growth regulating factor constraining biomass accumulation due to plant stress. These stresses include water, temperature, Nitrogen (N) and Phosphorus (P), calculated separately. The leaf area index (LAI), which is defined as the one-sided green leaf area per unit ground surface area, is then estimated by an empirical function of accumulated heat units and above-ground biomass. SWIM, therefore, produces yearly cumulative vegetation biomass (including for grasslands) as a function of these stressors.

6.3.1.2 Model calibration: SWIM

The hydrologic model was calibrated for 1990-1994 and validated for 1995-2000 using flow data at the catchment outlet. A static land-use map of 1990 reclassified from the National Land Cover Data of South Africa (South African Environmental Affairs, 1990) was used as a base and calibration map for the SWIM model. The five year calibration period was defined in such a way that it contains wet, average, and dry years. Parameters are calibrated manually by visual inspection of discharge plots and using performance indicators of the percent bias (PBIAS) (Moriasi et al., 2007) and the Nash-Sutcliffe efficiency (NSE) (Nash and Sutcliffe, 1970). PBIAS is an error index describing the average tendency of simulated values to be larger or smaller than observed data (Moriasi et al., 2007) as shown by Eq. 6-4.

$$PBIAS = \frac{\Sigma_j (Sim_j - Obs_j)}{\Sigma_j (Obs_j)} * 100\%$$ (6.4)

where Sim_j is the simulated and Obs_j is the observed value at time-step j, respectively.

In this case it is given in percent, and the optimal value is PBIAS = 0%, whereas model accuracy gets worse the greater the deviation from zero. Positive values indicate model overestimation bias and negative values indicate model underestimation bias. Nash-Sutcliffe efficiency (NSE) is a dimensionless model performance measure commonly used in hydrologic modelling. It is the ratio of residual variance to variance in observations, Eq. 6-5.

$$NSE = 1 - \frac{\sum_j (Sim_j - Obs_j)^2}{\sum_j (Obs_j - Obs_{mean})^2}$$ (6.5)

where Obs_{mean} is the mean of observations over the analysis's time period.

Model performances are generally considered satisfactory if NSE>0.5 and PBIAS< ±25% (Moriasi et al., 2007).

The land-use model

6.3.1.3 Inputs and setup

The land-use model was developed to simulate from 2000-2010. Input data for SITE included the initial land-use map of 2000 derived from GLC30- a 30m resolution global land cover product (GLC30, 2014; Jun et al., 2014). Land-use maps of 2000 and 2010 derived from this product are used for base and reference maps, respectively, for the land-use simulation model. The maps are reclassified and include land-use categories of water bodies, forest, shrub land, savanna, grassland, wetland, cultivated/cropland, vegetation mosaics (abbreviated as 'veg_mosaic' hereafter), urban settlement and bare land using two other high resolution (15m) local land-use products for the years 2005 and 2008 (Ezemvelo, 2011). Other inputs for the land-use model include data on population, livestock, biomass (average biomass dynamically passed from the hydrologic model at a yearly time-step), protected/reserved areas, computed distances from water, road, and urban centers and slopes derived from digital elevation models (DEMs). The model dynamically allocates land-uses based on projected demands derived by population and livestock for various uses (see section 6.3.2.3).

6.3.1.4 Model calibration: SITE

The SITE land-use change model was calibrated using the CALib genetic algorithm library integrated within SITE. The model performance was evaluated quantity disagreement (QD) and allocation disagreement (AD) measures (Pontius and Millones, 2011) (see also chapter 4). Starting simulation with the base map of 2000, the simulated map for the year 2010 is compared with the reference map for the same year using QD and AD measures. Unfortunately, no unique threshold exists in the literature that defines the acceptable values of accuracy for map comparisons in general. Although not particularly for QD and AD,

Landis and Koch (1977a) suggested agreement ranges of map comparison results into three groups: agreement value greater than 0.8 (80%) represents strong agreement; agreement value between 0.4 (40%) and 0.8 (80%) represent moderate agreement; and agreement value less than 0.4 (40%) represents poor agreement between two maps. We used this rough guide to evaluate the performance comparison between the simulated and the reference land-use maps.

6.3.1.5 Demand for land use

In relation to land-use demand projection for the land-use model, past trends of population, livestock, settlement, and land-use change are used. The population growth rate in the Thukela district is 0.17%/year (Lehohla, 2012). Increased livestock population in the rural community in this district, besides its commercial value, is a sign of more wealth and respect (Johnston et al., 2014; Salomon, 2006). The annual livestock growth rate for South Africa between 1990 and 2000 was reported to be 0.2%/year (FAO, 2004a, 2005). For the land-use change model simulation, we assumed the same annual livestock growth rate to continue to 2010. Therefore, annual population growth rate of 0.17% and livestock growth rate 0.2% together with their associated demands for various land uses and land-use related services are used for the model simulation.

6.3.1.6 Ecosystem services assessment

Grasslands provide ecosystem services in the form of, for instance, grazing, erosion control, water regulation, soil carbon retention, and biodiversity conservation (Lemaire et al., 2011; White et al., 2000). As a demonstration of the concept of integrated assessment modelling for ecosystem services assessment, and due to limitation of local datasets, only the grassland's grazing ecosystem service is quantified in this study. Thus we did not try to analyze comprehensive ecosystem service provisions by all land-use types, and neither did we try to analyze all ecosystem services provided by the grassland in the catchment. However, using a single, and major, ecosystem service associated with the grassland land-use, we demonstrate the effects of dynamic feedback between hydrology and land-use for ecosystem services assessment.

We used the concept of carrying capacity (Cowlishaw, 1969) to characterize the sustainability of grazing. Carrying capacity, with respect to livestock grazing, refers to the number of grazing animals a landscape is able to support without depleting rangeland vegetation or soil resources. It is defined as the area of land at a given time that is able to provide for a certain number of animals, expressed as animal-units per area (de Leeuw and Tothill, 1990; Tainton, 1999 ; Tainton et al., 1980). An animal unit (AU) is defined to be equivalent to a 450kg cow (Leistritz et al., 1992). One AU is assumed to consume 12kg of forage dry matter (biomass) per day, or 4.38 metric tons per year (Scarnecchia, 1985). Under sustainable management objectives, the actual amount of forage available for livestock grazing must be less than total biomass produced from the grassland (Fernández-Giménez and Swift, 2003). An adjustment for allowable use must be incorporated into the calculation to ensure that some un-grazed residual biomass is maintained to protect soil and vegetation resources (Kemp et al., 2000). With sustainable grassland ecosystems management in mind, the recommended minimum grazing capacity for the Thukela district is 0.5 AU/ha (Spehn et al., 2006). This would imply that at least a minimum of 6kg of dry matter per hectare per day (0.5*12kg/day), equivalent to 2.19 metric tons per hectare per year, should be available as residual biomass to maintain the sustainability of grazing ecosystem service of the grassland. Lower amounts imply overgrazing, soil erosion and/or land degradation on the catchment landscape which are deemed unsuitable for grazing. On studying factors influencing grassland grazing capacity, Holechek et al. (1995) suggest that distance from water and slope are also important considerations (see Tables 6-1 and 6-2).

Table 6-1. Reductions in grazing capacity with distance from water. Source: Holechek et al. (1995)

Distance from Water (km)	Reduction in Grazing Capacity (%)
0-1.6	0
1.6-3.2	50
>3.2	100

Table 6-2. Reductions in grazing capacity for different slopes. Source: Holechek et al. (1995)

Slope (%)	Reduction in Grazing Capacity (%)
0-10	0
11-30	30
31-60	60
>60	100

These distance and slope factors state that any amount of grassland biome outside or beyond the suggested upper limits of the ranges in these tables would imply that the resources are simply unreachable for livestock grazing. Integrating the above, the grazing capacity of a grassland ecosystem can be represented by Eq. 6-6:

$$G_c = B_y - \left(B_y * \frac{S_c G_c}{100} + B_y * \frac{D_c G_c}{100}\right) - B_{min} \tag{6.6}$$

where G_c=grazing capacity (in metric tons) per hectare per year; B_y=yearly grassland biomass (metric tons) per hectare per year; $S_c G_c$=percentage reduction in grazing capacity for the respective slope class; $D_c G_c$=percentage reduction in grazing capacity for the respective distance from water class; and B_{min}=minimum biomass to be maintained per hectare per year for a sustainable grassland ecosystem, i.e., 0.5AU/ha, which is equivalent to 2.19 metric tons/ha per year.

This equation (Eq. 6.6) is implemented on the land-use model for assessing the grazing capacity of the grassland using biomass produced and forwarded from SWIM at each time-step. Note that the grazing capacity G_c calculation produces only available biomass for livestock in terms of metric tons. This value can be divided by 2.19 to get the potential number of livestock the biomass can support. On the other hand, this potential livestock support can be compared with the density of livestock on the ground, whereby dynamic reduction of livestock may be implemented when enough forage is not available. However, given the high level (low resolution) spatial data about livestock density, we rather resorted to demonstrating the potential biomass capacity that can serve increasing livestock numbers (with annual rate of growth) on existing spatial units. So the assumption is that livestock is a dynamic component in the model (as it grows in number, its forage requirement increases too), but it is not dynamic in terms of, for example, response to shortage in forage supply.

The Model Coupling

While a universal definition for integrated modelling in environmental sciences is still evolving, it is generally accepted that it brings knowledge from two or more (sub-)domains to a common framework for interdisciplinary analysis. Integrated modelling involves linking of disciplines, processes, and/or scales depending on objective of the integration and data availability (Kelly et al., 2013). Model integration is often carried out through coupling of two or more specialized models from different (sub-) disciplines. Recently, a desire to integrate land-use change and hydrologic models has evolved (Karlsson et al., 2016; McColl and Aggett, 2007; Monier et al., 2016; Narasimhan et al., 2017). The benefit of such coupling is that watershed planners can examine the present and future characteristic of a specific watershed through analyzing effects of land-use change on water resources and vice-versa in an integrated and holistic manner.

6.3.1.7 Conceptual framework for the coupling

The conceptual framework of the coupling starts with development, calibration and evaluation of two separate land-use change (SITE) and hydrologic (SWIM) models independently. The two independent models are then coupled to exchange data between them dynamically. As shown in Figure 6-2, the SWIM model, which quantifies water availability for crop/vegetation growth, produces biomass for the land-use model (SITE). The land-use model determines (based on suitability factors and land-use demands) the land-use pattern which, together with the biomass output from SWIM, is used to quantify ecosystem services. Feedbacks from these can inform adaptation policies or further scenarios. Socio-economic, climatic and management scenarios are used as inputs on both models (Figure 6-2).

In the hydrologic model, the land-use map is provided dynamically from SITE. SWIM estimates surface runoff as a non-linear function of precipitation and a retention coefficient, which depends on soil water balance, land-use and soil type-a modification of the Soil Conservation Service Curve Number (SCS-CV) method (Krysanova and Wechsung, 2000; Mishra and Singh, 2013). From these sub-components of surface runoff, land-use attribute derived from the land-use map is dynamically imported to SWIM from the land-use model at the end of each year. Thus, changes in land-use (vegetation covers) produced using SITE affect curve number (CV) and leaf area index (LAI) parameters in the SWIM model.

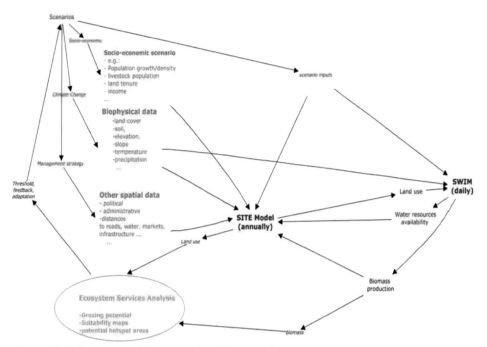

Figure 6-2. Conceptual framework of the coupling

In the land-use model, changes in land-use are determined following two procedures. First, land-use suitability template/map is produced based on soil type, soil texture, elevation, slope, and proximity from various resources and infrastructures, including water bodies, roads, and markets/urban centers. These factors are computed using multi-criteria analysis methods implemented in the suitability module in SITE. Then, based on the projected demand (associated with population and livestock growth rates) for the various land-uses, land-use is allocated on the template/suitability map. Both the uncoupled and the coupled models include these factors in their simulation of the land-use changes. The coupled model incorporates an additional factor, soil water balance, for computing land-use suitability. Soil water balance is dynamically imported from the hydrologic model. Thus, although some of these land-use suitability factors remain constant during the entire simulation period (such as elevation, slope, soil type, and soil texture, for example) others, including soil water balance, change dynamically. This dynamic continues until the end of the simulation period. In the meantime, the grassland's biomass (evaluated using the hydrologic model) and its potential for sustainable grazing ecosystem service are computed from the results of the coupling. Thus, soil water balance and aboveground biomass imported from the hydrologic model are used for computation of land-use suitability and for quantification of the grazing ecosystem service, respectively.

6.3.1.8 Technical details of the coupling

Like model integration, model coupling may mean different things for different people. Depending on the level of interaction between participating models, three forms of coupling are commonly reported in the literature: loose, tight, and embedded coupling (Bhatt et al., 2014; Brandmeyer and Karimi, 2000). Loose coupling involves exchange of results between two or more models with no need for modification within the participating models. No shared interface exists for the data exchange and thus participating models do not need to run in parallel (Bhatt et al., 2014). Tight coupling involves common interface or controlling unit and shared database for data exchange. Embedded coupling involves the merging (full integration) of processes and modules between participating models in a way that intra-model modification is possible in addition to shared database and a common user interface (Bhatt et al., 2014; Brandmeyer and Karimi, 2000). We developed a tight coupling between SWIM and SITE using the general framework presented in Figure 6-2. The coupling enables each of the two models to execute at a commonly defined time-step, which is one hydrologic year (October-September), using a common control script (interface) and a shared database (Figure 6-3).

The execution of the coupled models begins by initiating SWIM, using the control script, to run with the base land-use map input of year 2000. [Note that land-use map from 2000 is used as input to the hydrologic model for 2001, thus, the land-use from the end of the previous year is actually used as input to the hydrologic model in the beginning of the current year.] At the end of every yearly simulation, it modifies a log file to signal SITE to continue running using the results of SWIM.

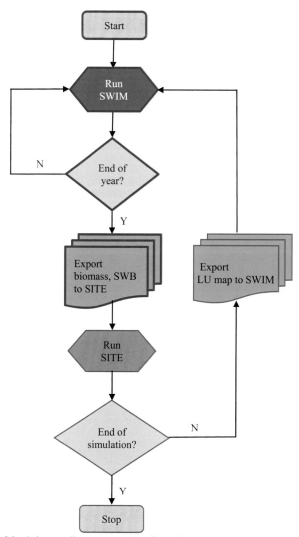

Figure 6-3. The Model coupling execution flowchart

[Note that N is used for 'no' and Y is used for 'yes' in the figure]

SITE executes its routine using these results for one time-step (1 yearly simulation) and
signals SWIM to continue with its simulation. This way, the Execute-Signal-Wait-Execute
cycle continues until both models come to the end of the defined simulation period for both.
Output of every time-step from one model is used as an input to the other model for the next
time-step (Figure 6-3).

Internal time-steps setup for each of the models are different: for the land-use model, one time-step is one year. Since the land-use model operates on a yearly time-step and the coupling data exchange is set to be at yearly time-step, the land-use model's 'internal' and 'external' simulation time-steps are both set at 1. Thus, output from the land-use model is passed to the hydrologic model at the end of each land-use simulation time-step. On the other hand, the hydrologic model operates on a daily basis, and thus has to use a single land-use map for 365 'internal' simulations. Land-use map of the year 2000, for instance, is used as input for the next 365 simulation time-steps (2001) in the hydrologic model. This tight coupling is handled with an external control script, 'coupler.py', that triggers and control the execution process in both models. Thus, after each 'external' simulation time-step, the two models read the execution status parameter in a parameter file , 'param.ini', from a shared database on whether to proceed with simulating or to wait for the other model to complete its time-step. This continues until both the land-use and the hydrologic models finish the simulation step that is defined in the same parameter file, 'param.ini'.

Software and scripts

In total, two existing and three newly developed software programs/scripts are used. The existing models are modified to facilitate the coupling and data exchange.

- **SWIM**: the source code of the FORTRAN based SWIM model was slightly modified for this coupling. Control routines to read/write model execution status, and recursively check for end of simulation step of the model so as to either proceed or wait, by reading from and writing to a commonly accessible file, are implemented in the main ('mainpro.f90') module of the SWIM model. Furthermore, after the first year of simulation, the grazing module in SWIM was modified to compute at the beginning of each new simulation time-step (after the whole grassland biomass has been forwarded to SITE) instead of at the end of each simulation time-step. This is done to facilitate full accounting of the grassland's carrying capacity in the land-use model before any reduction/grazing/regrowth module is applied in SWIM. Since the objective of the ecosystem services quantification is to assess the total sustainably graze-able grassland potential for livestock, the total grassland biomass is passed to the land-use model before reductions due to grazing.

- **SITE**: SITE was used for simulating land-use change and for implementing computation logics that quantify the grassland's potential for sustainable livestock grazing. The main application script in SITE has two functions: the Initialize() and SimulationStep() required by the system domain. The earlier initializes global variables, spatial references and initial grid-cell values. The 'SimulationStep()' function entails tasks to perform at each step of the land-use simulation. The functions to check model execution status, to import biomass and export land-use to/from the shared 'Exchange' data folder (Figure 6-3) are implemented in the 'SimulationStep()' function.

- **Ascii2Swim.py**: This python script was written to convert ASCII based grid cell land-use values from SITE and to recreate SWIM hydrotopes at each time-step based on these land-use values. It is discussed earlier that hydrotopes are dependent, among others, on land-use values. Whenever a new land-use map is imported to SWIM at each time-step, the hydrotopes inherently change with it. This script uses the GRASS GIS APIs (Application Programming Interfaces), which are functions for GIS processing in GRASS, for reclassification of hydrotopes and matching SWIM's sub-basins arrangement at each time-step.

- **Swim2Ascii.py**: This script was written to convert SWIM model outputs to an ASCII grid format for use by SITE. It uses GRASS API functions for the conversions at each time-step. Yearly average biomass values produced with SWIM are exported to the exchange folder in this format whereby SITE loads them for its own computations at each time-step.

- **Coupler.py**: This was written to initiate parallel executions of SITE and SWIM, and manage execution processes, irregularities (such as missing input files, delayed response) or errors during execution of the coupled models. It lets the two models run in parallel and proceed with their own input, processing and output unless irregularities/errors in execution are reported.

It is to be noted that only newly developed scripts or modified models that are related to the coupling are listed here. Other dependency tools (such as pyWin, wxWidgets) and pre-processing tools such as GRASS and ArcGIS are used as well.

Operating environment

Although the individual models can run on Linux operating systems, the coupled models are developed and tested in a Windows environment only. Specifically, windows 7, 64 bit machines are used. The Photran integrated development environment (IDE), based on Eclipse and CDT (C/C++ development tooling), is used for compiling the FORTRAN based modified SWIM code on Eclipse Luna (2016).

6.4   Results

Results corresponding to land-use change and hydrologic components are analyzed both with and without the coupling of the two respective models. Furthermore, the grassland's potential for sustainable grazing ecosystem service and future scenarios are assessed. The results of this assessment are presented in the following subsections.

Hydrologic changes

The uncoupled SWAT model was calibrated for 1990-1994 and evaluated for 1995-2000 (see Table 6-3 for calibration parameters). Performance measures show NSE values of 0.58 and 0.53, and PBIAS values of -0.17 and -0.21 for the calibration and evaluation periods of the model, respectively (see Figure 6-4 and Table 6-4).

Table 6-3. Calibrated model parameters and their final values

Parameters*	*ecal thc roc2 roc4 bff sccor abf delay revapc rchrgc revapmn*
Values	0.53 1.0 0 0.51 1 1.33 0.1 100 0.21 0.04 0

*ecal->correction factor for potential evapotranspiration; **thc** ->correction factor for sky emissivity-affects potential evapo-transpiration; *roc2* ->routing coefficient -storage time constant of surface flow; *roc4* ->routing coefficient -storage time constant of subsurface flow; *sccor* ->correction factor for saturated conductivity; **bff** ->baseflow factor; **abf** ->alpha factor for groundwater; *delay*->groundwater delay (days); **revapc** ->fraction of groundwater recharge that evaporates; *rchrgc* ->fraction of shallow groundwater that percolates to deep; **revapmn**->threshold of groundwater storage before evaporation can start (mm).

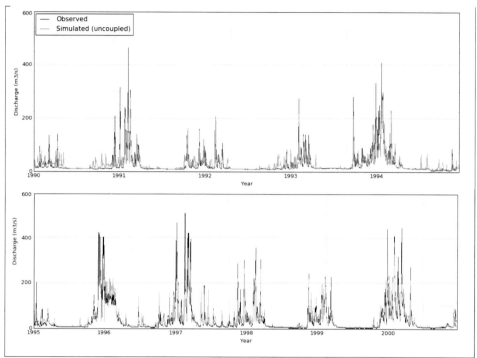

Figure 6-4. Calibration and validation of the uncoupled hydrologic model

Table 6-4. Coupled vs. uncoupled SWIM model performance against observed streamflow

Model	Setup	Calibration (1990-1994)		Validation (1995-2000)		Validation (2001-2010)	
		NSE	PBIAS	NSE	PBIAS	NSE	PBIAS
SWIM	Uncoupled	0.58	-0.17	0.53	-0.21	0.52	-0.22
	Coupled					0.54	-0.19

Then, SWIM was run from 2001 to 2010 both in coupled and uncoupled mode. Both the coupled and the uncoupled models are re-evaluated against the observed data from 2001-2010 (Figure 6-5 and 6-6). The uncoupled SWIM model showed performance values of 0.52 and -0.22 for NSE and PBIAS measurements, respectively. The coupled hydrologic model showed values of 0.54 and -0.19 for NSE and PBIAS, respectively. Note that SWIM was not re-calibrated for 2001-2010; the simulated dataset is simply re-evaluated against the observed dataset in terms of NSE and PBIAS measures.

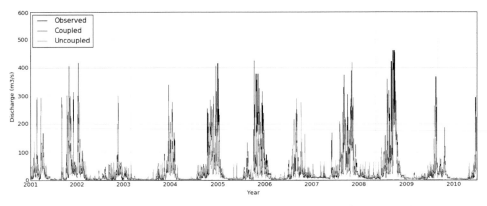

Figure 6-5. Comparison of coupled, uncoupled and observed flow 2001-2010

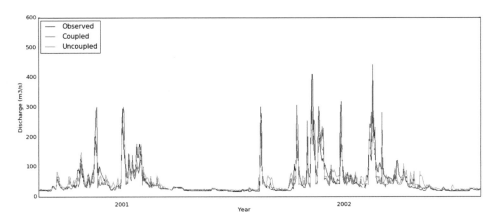

Figure 6-6. A closer look of the coupled and uncoupled flow simulations (2001-2002)

Table 6-5. Overall average comparison with observed flow (% difference) from 2001-2010

Flow components*	Model	
	Coupled model	Uncoupled model
High-flow	-18.7	-26.5
Average-flow	-5.2	-7.0
Low-flow	-6.3	-4.5

*N.B. High-flow is represented by flows of 200 m^3/s and above, average-flow is represented by flows from 30 m^3/s to 200 m^3/s, and low-flow is represented by flows below 30 m^3/s, all assumed from observation of the streamflow hydrograph.

Overall average flow comparison in terms of percentage difference with the observation is shown in Table 6-5. For this analysis, the stream-flow hydrograph was segmented in regimes of high-flow, average-flow and low-flow regimes/seasons via visual examination of the streamflow hydrograph in Figure 6-5. Accordingly, flows above 200 m^3/s are assumed as high-flows, and those in between 30 m^3/s and 200 m^3/s are assumed to be average-flows. Flows less than 30 m^3/s are assumed to be low-flows for this analysis. From Table 6-5, we can see that the result of the simulated flow from the uncoupled model has a larger percentage difference against the observed flow on high-flow seasons compared with the result from the coupled model. Results of the average-flow and the low-flow seasons from both the uncoupled and the coupled models are comparable, with a slight advantage of the uncoupled on the low-flows and the coupled on the average-flows.

Land-use change

The land-use model was simulated from 2001 to 2030. Both coupled and uncoupled simulated maps for the year 2010 are evaluated against the reference land-use map for the same year. Evaluation of the coupled simulation map against the reference land-use map for 2010 showed a QD of 6.4% and an AD of 8.1%, adding up to a total disagreement of 14.5% (total agreement of 85.5%) between the two land-use maps (Table 6-6). From this table, evaluation of the uncoupled simulation against the reference land-use map for the same year showed a QD of 7.5% and an AD of 8.7%, adding up to a total disagreement of 16.2% (total agreement of 83.8%).

Table 6-6. Comparison of simulated vs reference maps (GLC30) of 2010

Model	Quantity and allocation disagreement			
	Components			Measure (%)
	Change simulated as 'persistence' (QD)	Persistence simulated as 'change' (QD)	Change simulated as 'change to wrong category' (AD)	Total disagreement
Coupled	2.9	3.5	8.1	14.5
Uncoupled	3.2	4.3	8.7	16.2

Results from two levels of dynamics are analysed within the land-use model using the methodologies described thus far. In the first level, the changes in land-use, simulated using the SITE land-use change model, are observed (Figure 6-7and Figure 6-8).

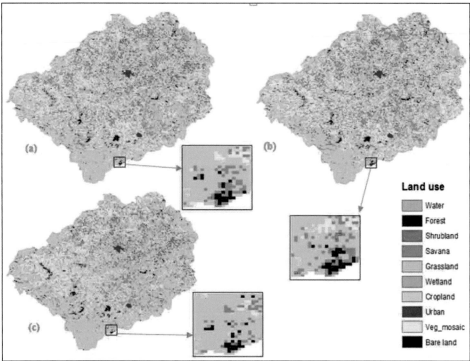

Figure 6-7. Reference map of 2010 (a), and simulated maps of 2010 from the uncoupled (b) and the coupled (c) models.

The second level of dynamics is the changes in sustainable grazing capacity of the grassland cover of the catchment (Figure 6-9). Simulated land-use map for the year 2010 is shown on Figure 6-8 together with reference land-use maps of 1990 and 2000. As it can be seen from the figure, land-use change trends show increases in 'Cropland' and 'Veg_mosaic' and decreases in 'Grassland', 'Forest', 'Shrubland' and 'Savana'.

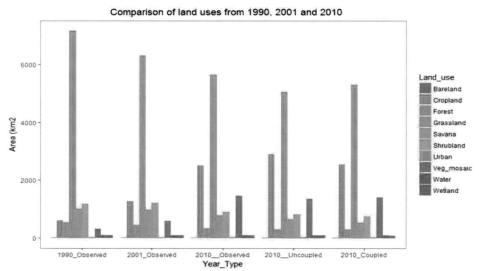

Figure 6-8. Land cover changes: 1990-2010

Quantification of the grassland's grazing ecosystem service in the Thukela catchment using both the coupled and the uncoupled models is shown in Figure 6-9. The figure shows that grassland biomass and the associated grazing ecosystem service in the catchment are in decreasing trend in general. The coupled model shows higher amount of biomass but lower value of grazing ecosystem service, whereas the uncoupled model shows lower amount of biomass and yet higher amount of grazing ecosystem services.

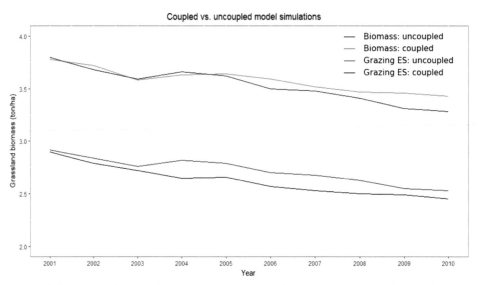

Figure 6-9. Trends in grassland biomass and grazing ecosystem services using the coupled and the uncoupled models

6.5 Discussion

From the perspective of hydrologic modelling, the calibration and evaluation values for the uncoupled model show NSE measures of 0.58, 0.53, and 0.52 for the calibration (1990-1994) and validation (1995-2000, 2001-2010) periods, respectively. A model performance with NSE value >0.5 is often taken as satisfactory in hydrologic modelling (Moriasi et al., 2007). Accordingly, the performance of the model simulation was deemed acceptable. Likewise, PBIAS measures show values -0.17, -0.21, and -0.22 for the calibration and validation (1995-2000, 2001-2010) periods in the uncoupled hydrologic model, respectively (Table 6-4). Negative PBIAS values show that the simulated model generally underestimates the observed flow. However, the values are within the acceptable range (within ±25%) for advisable PBIAS measure (Moriasi et al., 2007). Evaluation of the coupled models (2001-2010) showed NSE and PBIAS performance values of 0.54 and -0.19, respectively (Table 6-4). Both the NSE and the PBIAS values for the coupled and the uncoupled models are relatively close, with slight improvement in favour of the coupled model in both measures.

A closer look between results of the coupled and the uncoupled hydrologic models for the period between 2001-2010 (Figure 6-6) show that the coupled model captures the observed flow, especially during high-flow seasons, better than the uncoupled model. This is shown to be the case using results (shown in Table 6-5) derived from the flow hydrograph. The analysis shows that, when compared with the observed flow, the coupled model has an overall smaller percentage difference on average during high-flow seasons whereas it is relatively comparable with the uncoupled model during average and low-flow seasons (see Table 6-5). Better performance of the coupled model may be because the dynamic changes in land-use better reflect the actual situation, leading to a better simulation of the runoff generation process. The overall difference between the coupled and the uncoupled hydrologic models, as can also be seen from the performance measures in Table 6-4, seems marginal, however.

With regards to changes in land-use, a trend of decrease in grassland and increase in cropland can be observed from base and reference maps of 1990, 2000 and 2010 (Figures 6-7 and 6-8). Both the coupled and the uncoupled land-use simulation results of 2010 are compared with the reference map of 2010 (GLC30). The overall performance difference, as measured using the quantity and allocation disagreements, between the coupled and the uncoupled models, 14.5% and 16.2%, respectively, is modest, and not as significant as we would expect. Results from both the coupled and uncoupled land-use change models for 2010 (Figure 6-8) show notable differences in the simulation of the land-use change trend. In the uncoupled model, lower 'grassland' and higher 'cropland' allocation is shown compared to the ones in the coupled model for the same year. This is due to the fact that, in the uncoupled model, no restriction or factor is set with regards to availability of water for crop suitability. Thus, more land-use pixels (including grassland pixels) that may normally be less suitable in terms of water availability will get converted and allocated to cropland due to lack of information on water dynamics. This results in allocation of more grassland pixels to cropland in response to higher demands for the later. In the coupled version, the land-use change model gets more accurate water related constraints to compute land-use suitability for land-use allocation. Thus the coupled land-use change model constrains the suitability of pixels for cropland, for instance, on which water availability is limited. In regards to the grassland allocation, the coupled model is more optimistic on the catchment's suitability for grassland (based also on water availability input) when compared to the uncoupled model (which does not consider such an input). On the other hand, the uncoupled model allocates more land uses in response to cropland demands irrespective of water availability, i.e., due to lack of water availability information. Comparison of the coupled and the uncoupled land-use models show that hydrologic components/water availability can constrain suitability of a catchment for various land-use purposes (Figure 6-7 and 6-8). Furthermore, the coupled versions of both the hydrologic as well as the land-use change models show modest improvement in performance when compared to their uncoupled counterparts. However, the difference between the performances could well be argued to be marginal, and thus difficult to unequivocally separate signal from noise from these results.

Unlike the marginal improvement on performances between the coupled and the uncoupled versions in the respective models, the difference in quantification of ecosystem services assessment (of grazing in this case) between results of the coupled and the uncoupled models is significant, however. As we can see from Figure 6-9, the difference between the coupled and the uncoupled model results is attributed to lack of water availability information in the uncoupled model. Without feedback from the hydrologic model on the dynamics of water availability, the grassland vegetation growth could not be constrained for water demands and climatic variables. Areal coverage of a grassland from the land-use model alone would not have been enough to explicitly estimate the sustainability of the grazing ecosystem service as presented here. This is important for informed decision making on integrated management of natural resources in general and for spatially explicit quantification of ecosystem services catchment ecosystem services in particular.

In summary, the study shows that dynamic feedback between land-use and hydrology improves model performances only marginally. The trends both models follow were also slightly different, where the coupled model was especially better at capturing high-flows in the hydrologic model. The coupled land-use model showed slightly better performance than the uncoupled land-use model. The significance of the model coupling, however, was shown much better through simulating dynamic feedback between the two models for assessment of the sustainability of the grazing ecosystem service potential of the grassland in the Thukela catchment. As shown earlier, availability of water together with climate variables (for suitability assessment in the land-use model) as well as yearly average biomass (for computation of the grazing ecosystem services) could not have been analysed from either of the uncoupled models alone. Thus, besides filling the conceptual gaps in land-use and hydrologic model representations argued from the outset in this chapter, model coupling provides additional potential for exploration and adds a new dimension for assessment of environmental problems (such as the ecosystem service assessment in this study) that may not be addressed from individual models alone. To the best of our knowledge, this practice is nearly non-existent, or non-reported, and only static land-use maps, from a single episode or a couple of periodical episodes, are oftentimes used to analyse land-use impacts on hydrology. Furthermore, although the magnitude of the overall land-use change impact on the hydrology seems less pronounced, a closer look at the sub-components of the flow hydrograph has shown that sub-components of the streamflow responded differently to the dynamics of the land-use (see Table 6-5). Likewise, hydrology is only barely, if ever, represented in land-use change models, often through the use of either precipitation or through analysis of proxies such as distance to water sources. Such practices in modelling in general, we argue, have downplayed the reported effect of the dynamics of land-use change in hydrology and vice-versa, and even more so for the assessment of ecosystem services dependent on these interacting domains. In that regard, we believe that this study establishes an opportunity for further research in the area of ecosystem services, a relatively young domain, within the framework of coupled models particularly from established domains of hydrology and land-use change. This being said, there is ample scope for improving this study. First, besides model uncertainties, lack of high resolution and reliable data especially on socio-economic inputs will have influenced model results. Second, as is usually the case with integrated models, coupling of the two models involves several parameters, tools and techniques. The number of tools involved, the need for modifying both models at code level and a number of spatial and temporal parameters on both (such as simulation time-steps and spatial resolutions) can make model coupling technically demanding. Alternative coupling methods, for instance for pre-defined input and outputs such as OpenMI (Gregersen et al., 2007), may be consulted in cases where less technical endeavours in data exchange, but predefined or relatively rigid, interaction between two models is desired. In spite of these limitations, however, we have noted that regardless of magnitudes, dynamic feedback of land-use change affects hydrologic response, and that dynamic feedback of hydrology affects land-use responses. Even more so, results of dynamic feedback between hydrology and land-use change affect quantification of ecosystem services in a catchment. Hence, a deserving attention should be given to the dynamic interaction of land-use and hydrology during the development of modelling concepts or modelling activities in respective models for a more accurate and explicit quantification of catchment ecosystem services.

6.6   Conclusions

This chapter presented an analysis of the effects of dynamic feedbacks between land-use and hydrology for the quantification of ecosystem services using an integrated modelling approach. The model integration/coupling exchanges annual aboveground biomass from the hydrologic model to the land-use model and land-use maps from the land-use to the hydrologic model. Results show that dynamic feedback between the hydrologic and the land-use change models show marginal improvements in performance in both models (for SWIM, NSE and PBIAS values barely increased and for SITE the map-comparison statistics improved only slightly). Thus, it can be argued that coupling land-use change and hydrologic models may not be always necessary for general modelling purposes. However, the effect of dynamic feedback between hydrology and land-use change was shown more clearly to affect quantification of ecosystem services in this study. A serious modelling effort for specific needs (such as to analyse effects of land-use change on runoff generation for purposes of flooding/ peak-flow, etc.) or for quantifying catchment ecosystem services associated with land and water resources would be advised to account for the dynamic feedback between these domains. Accounting for the dynamic feedback can also serve to fill the conceptual gaps in representation of the interactions between the respective models observed in the literature. For a general practice, however, it can be argued that model integration, and thus model coupling, increases technical complexity. Thus, modellers and environmental decision makers should weigh between the need for a better performance and complexity when it comes to practical applications of coupling.

Chapter 7. Conclusions and Recommendations

7.1 Conclusions

Interactions of land-use and hydrological processes are gaining increasing interest with the growing endeavor for a more sustainable and integrated natural resources management. The intensity, rate and magnitude of land-use change influence hydrological responses, which may influence further changes in land use. These two phenomena are dynamically linked. Sustainable management of land and water resources requires an integrated assessment of these resources and benefits from spatially explicit and integrated models and frameworks. Such models and frameworks are able to represent in a simplified way the availability and suitability of land or water resources and their impacts on each other for local or regional analysis and decision making purposes. This research adopted an integrated modelling approach for the analysis of various modelling experiments on land and water resources. As such, this research developed methods and tools for identifying and ranking land-use change drivers, for developing, parameterizing and simulating land-use change models, for assessing dynamic land-use suitability using global tools and datasets as well as for analyzing the effects of dynamic feedback between land use and hydrology.

Land-use change modelling

Land-use change modelling involves complex layers of socio-economic and biophysical factors. With the objective of developing a predictive land-use model, we analyzed socio-economic and biophysical land-use drivers. We also developed a land-use change model that was parameterized and calibrated using field data. Using an initial land cover of 1986 as a base map, we developed and simulated a land-use model until 2009. After calibration and evaluation, the simulated map of the 2009 land cover was compared with the reference (Landsat derived) land cover for the same year. The simulated model was further continued to simulate to the year 2025 under a business-as-usual scenario. This scenario assumes present rates of growth in population and livestock as well as associated demands to continue the same. The simulated map showed an overall good performance in mimicking trends and magnitudes of the observed land cover map. It was concluded that field based parameterization and detailed representation of socio-environmental land-use change drivers were imperative for an accurate representation of catchment land-use dynamics.

Land suitability evaluation

Land-use change models commonly use, among other things, proximity factors such as distances from various resources and infrastructures such as roads, water bodies and markets to determine land-use suitability. Based on such inputs, they compute land-use suitability at the initial time step of the model simulation and allocate land-uses based on this initial suitability map. However, such inputs are often dynamic: new roads, new markets and/or new infrastructure would mean a change in land-use suitability. To assess this fact, dynamic land-use suitability was analyzed using various integrated assessment methods and with a focused analysis on the most dominant land-use type (agricultural land) in the Abbay basin. Results of the land suitability assessment show that, even though unused suitable lands are available in the basin, marginal lands are also being used for agriculture especially in the highlands of the basin, which might imply an aggravated land degradation and soil erosion. It was also noted that freely available global datasets and computing platforms can be very useful for integrated natural resources management and assessments including land-suitability assessment.

Feedback between land-use and hydrology

The way the interactions between land-use change and hydrologic components are represented in many models is often problematic. Most land-use models do not take account of hydrologic feedbacks, whereas hydrologic models often assume a static land-use during model simulation. To address impacts of the two-way feedback between land-use and hydrology, two separately developed land-use and hydrologic models were coupled to dynamically exchange outputs of one as input to the other. The impact of dynamic feedback on land-use change was demonstrated in terms of changes in quantity of the grassland ecosystem services for grazing in the Thukela catchment, South Africa. Hydrologic response to land-use change was demonstrated by assessing changes in river flow at the end of the catchment outlet of the Thukela. Noticeable changes in the amount of river discharge especially during high flow periods due to changes in land-use lead to the conclusion that hydrologic modelers who model catchment areas with significant changes of LULC should seriously consider incorporating dynamic land-use maps in their models especially if their application focuses on high-flow related analysis such as for flood forecasting purposes. Likewise, hydrologic processes and components resulted in noticeable impacts on land-use change and thereby on the quantity and quality of ecosystem services. It can be concluded that the use of two-way and dynamic feedback between land-use and hydrologic models produces improved information that can better inform sustainable and integrated natural resources management efforts.

Web-based framework based on GEE

Massive amounts of socio-environmental data are added to global data archives every year. The importance of these datasets in decision and policy support for integrated natural resources management is evident. However, the use of such large amounts of global datasets for local and regional decision making on integrated land and water resources assessment has been limited due to needs for efficient access, storage and sufficient computational power. The Google Earth Engine (GEE) has presented an opportunity that can address these limitations. With a web-based and automated implementation of dynamic land-use suitability analysis on the Abbay basin, it is shown computational power that are made available through platforms such as GEE can be used to resolve 'Big Data' bottlenecks in environmental assessment. Access, storage and computational power that are increasingly made available can be harnessed to address challenges related to local and regional environmental problems.

In general, this study concludes that:

1. Catchment ecosystem services are inseparably linked with land use and water resources. Quantification of these services requires a holistic and integrated modelling framework.

2. Land-use suitability involves multiple socio-economic and biophysical criteria. Although some of these criteria have been universalized though various studies, local criteria and specifics determine the accurateness of these assessments.

3. For an accurate representation of land-use change dynamics in catchments, a thorough assessment and representation of both socio-economic and biophysical land-use change drivers and incorporating these drivers in land-use change models is important.

4. Including dynamic feedback between hydrologic and land-use change models consistently improves model performance. It is also a logical improvement to existing models and modelling frameworks. In practice, however, this process is data intensive and increases model complexity, which may be prohibitive for general purpose modelling.

5. Global datasets and tools can provide easy access to data and computing resources for generating valuable information on natural resources especially in data scarce catchments in Africa. However, results based on such datasets are still less reliable for local decision support purposes without thorough evaluations using local knowledge and ground data.

7.2 Recommendations

Integrated assessment modelling involves a number of scientific domains and tools that are interconnected to produce a holistic view of environmental systems. Integrated assessment and modelling effort carried out on land and water resources in this research revealed a number of challenges and opportunities for further research. These challenges and opportunities range from evaluation and calibration of land-use change models to complexities involving model coupling from various domains. The following recommendations are outlined for further research with respect to integrated land and water resources assessment:

1. Investigating the effects of dynamic feedback between land-use and hydrology on trade-off and synergy between multiple catchment ecosystem services, such as trade-offs between grassland ecosystem services for livestock grazing and for water regulation, for example.

2. Investigating the explicit role and implication of climate change on the feedback between land-use change and hydrology.

7.3 Limitations

Major limitations of this study are:

1. The study developed and tested models and tools in data scarce catchments. As such, evaluation and calibration performance of the individual models involved in the study, particularly that of hydrology, was not very high. Detailed hydrologic modeling and analysis of the catchment will require further fine-tuning of the models with measured parameters and more complete datasets.

2. As is often the case, integrated modeling involves multiple models and tools. Besides lack of reliable data, this unavoidably introduces model uncertainties. Unfortunately, such uncertainties are difficult to quantify in integrated modeling in general, and were not quantified in this study.

124

REFERENCES

Abdi, H., 2003. Factor rotations in factor analyses. Encyclopedia for Research Methods for the Social Sciences. Sage: Thousand Oaks, CA, 792-795.

Abdi, H., Williams, L.J., 2010. Principal component analysis. Wiley Interdisciplinary Reviews: Computational Statistics 2, 433-459.

Abera, W., Formetta, G., Brocca, L., Rigon, R., 2017. Modeling the water budget of the Upper Blue Nile basin using the JGrass-NewAge model system and satellite data. Hydrology and Earth System Sciences 21, 3145.

Akıncı, H., Özalp, A.Y., Turgut, B., 2013. Agricultural land use suitability analysis using GIS and AHP technique. Computers and Electronics in Agriculture 97, 71-82.

Ali, Y.S., Crosato, A., Mohamed, Y.A., Wright, N.G., Roelvink, J., 2014. Water resource assessment along the Blue Nile River, north Africa with a one-dimensional model. Proceedings of the Institution of Civil Engineers 167, 394.

Altman, D., 1991. Comparing groups—categorical data. Practical statistics for medical research 1, 261-265.

Amézquita, E., Preciado, G., Arias, D., Thomas, R., Friesen, D., Sanz, J., 1998. Soil physical characteristics under different land use systems and duration on the Colombian savannas, World Congress of Soil Science XVI, Montpellier, France.

Amouroux, E., Chu, T.-Q., Boucher, A., Drogoul, A., 2009. GAMA: an environment for implementing and running spatially explicit multi-agent simulations, Agent computing and multi-agent systems. Springer, pp. 359-371.

Anane, M., Bouziri, L., Limam, A., Jellali, S., 2012. Ranking suitable sites for irrigation with reclaimed water in the Nabeul-Hammamet region (Tunisia) using GIS and AHP-multicriteria decision analysis. Resources, Conservation and Recycling 65, 36-46.

Arnold, J.G., Srinivasan, R., Muttiah, R.S., Williams, J.R., 1998. Large area hydrologic modeling and assessment part I: Model development1. Wiley Online Library.

Ashagrie, A., De Laat, P., De Wit, M., Tu, M., Uhlenbrook, S., 2006. Detecting the influence of land use changes on discharges and floods in the Meuse River Basin: the predictive power of a ninety-year rainfall-runoff relation. Hydrol. Earth Syst. Sci. 10, 691-701.

Asner, G.P., Elmore, A.J., Olander, L.P., Martin, R.E., Harris, A.T., 2004. Grazing systems, ecosystem responses, and global change. Annu. Rev. Environ. Resour. 29, 261-299.

Aspinall, R., Justice, C., 2003. Land use and land cover change science strategy, Summary of a workshop help at the Smithsonian Institute. Organized by the US Climate Change Science Program-CCSP Land Use Interagency Working Group-LUIWG. Nov, p. 20.

Asres, M.T., Awulachew, S.B., 2010. SWAT based runoff and sediment yield modelling: a case study of the Gumera watershed in the Blue Nile basin. Ecohydrol Hydrobiol 10, 191-199.

Asres, S.B., 2016. Evaluating and enhancing irrigation water management in the upper Blue Nile basin, Ethiopia: The case of Koga large scale irrigation scheme. Agricultural Water Management 170, 26-35.

Awulachew, S.B., Ahmed, A., Haileselassie, A., Yilma, A., Bashar, K., McCartney, M., Steenhuis, T., 2010. Improved water and land management in the Ethiopian highlands and its impact on downstream stakeholders dependent on the Blue Nile.

Awulachew, S.B., McCartney, M., Steenhuis, T.S., Ahmed, A.A., 2009. A review of hydrology, sediment and water resource use in the Blue Nile Basin. IWMI.

Ball, G.H., Hall, D.J., 1965. ISODATA, a novel method of data analysis and pattern classification. DTIC Document.

Bandyopadhyay, S., Jaiswal, R., Hegde, V., Jayaraman, V., 2009. Assessment of land suitability potentials for agriculture using a remote sensing and GIS based approach. International Journal of Remote Sensing 30, 879-895.

Bashar, K., Zaki, A., 2005. SMA Based Continuous Hydrologic Simulation of the Blue Nile, International Conference of UNESCO Flanders, FUST.

Bel, L., Allard, D., Laurent, J., Cheddadi, R., Bar-Hen, A., 2009. CART algorithm for spatial data: Application to environmental and ecological data. Computational Statistics & Data Analysis 53, 3082-3093.

Betrie, G., Mohamed, Y., Griensven, A.v., Srinivasan, R., 2011. Sediment management modelling in the Blue Nile Basin using SWAT model. Hydrology and Earth System Sciences 15, 807-818.

Bewket, W., 2002. Land cover dynamics since the 1950s in Chemoga watershed, Blue Nile Basin, Ethiopia. Mountain Research and Development 22, 263-269.

Bewket, W., Abebe, S., 2013. Land-use and land-cover change and its environmental implications in a tropical highland watershed, Ethiopia. International journal of environmental studies 70, 126-139.

Bewket, W., Sterk, G., 2002. Farmers' participation in soil and water conservation activities in the Chemoga watershed, Blue Nile basin, Ethiopia. Land Degradation & Development 13, 189-200.

Bewket, W., Sterk, G., 2005. Dynamics in land cover and its effect on stream flow in the Chemoga watershed, Blue Nile basin, Ethiopia. Hydrological Processes 19, 445-458.

Bewket, W., Teferi, E., 2009. Assessment of soil erosion hazard and prioritization for treatment at the watershed level: case study in the Chemoga watershed, Blue Nile Basin, Ethiopia. Land Degradation & Development 20, 609-622.

Bhaduri, B., Harbor, J., Engel, B., Grove, M., 2000. Assessing watershed-scale, long-term hydrologic impacts of land-use change using a GIS-NPS model. Environ Manag 26, 643-658.

Bhatt, G., Kumar, M., Duffy, C.J., 2014. A tightly coupled GIS and distributed hydrologic modeling framework. Environ. Model. Software 62, 70-84.

Bitew, M., Gebremichael, M., 2011. Assessment of satellite rainfall products for streamflow simulation in medium watersheds of the Ethiopian highlands. Hydrology and Earth System Sciences 15, 1147-1155.

Bizuwerk, A., Peden, D., Taddese, G., Getahun, Y., 2005. GIS Application for analysis of Land Suitability and Determination of Grazing Pressure in Upland of the Awash River Basin, Ethiopia. International Livestock Research Institute (ILRI), Addis Ababa, Ethiopia.

Bojórquez-Tapia, L.A., Diaz-Mondragon, S., Ezcurra, E., 2001. GIS-based approach for participatory decision making and land suitability assessment. International Journal of Geographical Information Science 15, 129-151.

Bonfante, A., Bouma, J., 2015. The role of soil series in quantitative land evaluation when expressing effects of climate change and crop breeding on future land use. Geoderma 259, 187-195.

Bontemps, S., Defourny, P., Bogaert, E.V., Arino, O., Kalogirou, V., Perez, J.R., 2011. GLOBCOVER 2009-Products description and validation report.

Boroushaki, S., Malczewski, J., 2008. Implementing an extension of the analytical hierarchy process using ordered weighted averaging operators with fuzzy quantifiers in ArcGIS. Computers & Geosciences 34, 399-410.

Boyd, J., Banzhaf, S., 2007. What are ecosystem services? The need for standardized environmental accounting units. Ecological Economics 63, 616-626.

Brandmeyer, J.E., Karimi, H.A., 2000. Coupling methodologies for environmental models. Environ. Model. Software 15, 479-488.

Breuer, L., Eckhardt, K., Frede, H.-G., 2003. Plant parameter values for models in temperate climates. Ecol Model 169, 237-293.

Brinkman, R., Young, A., 1976. A framework for land evaluation. FAO.

Brown, D.G., Walker, R., Manson, S., Seto, K., 2004. Modeling land use and land cover change, Land Change Science. Springer, pp. 395-409.

Calder, I.R., 1998. Water-resource and land-use issues. Iwmi.

Cengiz, T., Akbulak, C., 2009. Application of analytical hierarchy process and geographic information systems in land-use suitability evaluation: a case study of Dümrek village (Çanakkale, Turkey). International Journal of Sustainable Development & World Ecology 16, 286-294.

Chakhar, S., Mousseau, V., 2008. Multicriteria spatial decision support systems, Encyclopedia of GIS. Springer, pp. 753-758.

Chamberlin, J., Tadesse, M., Benson, T., Zakaria, S., 2007. An Atlas of the Ethiopian Rural Economy: expanding the range of available information for development planning. Information development 23, 181-192.

Chen, H., Pontius Jr, R.G., 2010. Diagnostic tools to evaluate a spatial land change projection along a gradient of an explanatory variable. Landscape Ecology 25, 1319-1331.

Clarke, K.C., 2008. Mapping and modelling land use change: an application of the SLEUTH model, Landscape Analysis and Visualisation. Springer, pp. 353-366.

Costa, M.H., Botta, A., Cardille, J.A., 2003. Effects of large-scale changes in land cover on the discharge of the Tocantins River, Southeastern Amazonia. Journal of Hydrology 283, 206-217.

Cowlishaw, S., 1969. The carrying capacity of pastures. Grass Forage Sci 24, 207-214.

CSA, 2007. Population and Housing Census of Ethiopia. Central Statistical Agency, Addis Ababa, Ethiopia, p. 83.

CSA, 2008. Summary and statistical report of the 2007 population and housing census. Federal Democratic Republic of Ethiopia, Addis Ababa.

Dang, A.N., Kawasaki, A., 2017. Integrating biophysical and socio-economic factors for land-use and land-cover change projection in agricultural economic regions. Ecol Model 344, 29-37.

Das, S., Priess, J.A., Schweitzer, C., 2012. Modelling regional scale biofuel scenarios–a case study for India. GCB Bioenergy 4, 176-192.

de Leeuw, P.N., Tothill, J.C., 1990. The concept of rangeland carrying capacity in sub-Saharan Africa: Myth or reality. Overseas Development Institute, Pastoral Development Network London.

Decagon Devices, 2016. Plant Available Water: How do I determine Field capacity and permanent wilting point? [Online] 2016. [Cited: December, 2016.] http://www.decagon.com/support/datatrac-3-online-help-files/how-do-i-graph-plant-available-water/plant-available-water-how-do-i-determine-field-capacity-and-permanent-wilting-point/.

DeFries, R., Eshleman, K., 2004. Land-use change and hydrologic processes: a major focus for the future. Hydrological Processes 18, 2183-2186.

DeFries, R.S., Foley, J.A., Asner, G.P., 2004. Land-use choices: balancing human needs and ecosystem function. Frontiers in Ecology and the Environment 2, 249-257.

Dercon, S., Hoddinott, J., 2004. The Ethiopian rural household surveys: Introduction. International Food Policy Research Institute, Washington, DC Photocopy.

Desta, L., 2000. Land degradation and strategies for sustainable development in the Ethiopian highlands: Amhara Region. ILRI (aka ILCA and ILRAD).

Di Baldassarre, G., Viglione, A., Carr, G., Kuil, L., Yan, K., Brandimarte, L., Blöschl, G., 2015. Debates—Perspectives on socio-hydrology: Capturing feedbacks between physical and social processes. Water Resources Res. 51, 4770-4781.

Di Gregorio, A., 2005. Land cover classification system: classification concepts and user manual: LCCS. Food & Agriculture Org.

Dietterich, T.G., 1998. Approximate statistical tests for comparing supervised classification learning algorithms. Neural computation 10, 1895-1923.

Dingman, S., 2002. Water in soils: infiltration and redistribution. Physical hydrology. upper saddle river, New Jersey: Prentice-Hall, Inc.

Dominati, E., Mackay, A., Bouma, J., Green, S., 2016. An Ecosystems Approach to Quantify Soil Performance for Multiple Outcomes: The Future of Land Evaluation? Soil Science Society of America Journal.

Du, X., Jin, X., Yang, X., Yang, X., Zhou, Y., 2014. Spatial pattern of land use change and its driving force in Jiangsu province. International journal of environmental research and public health 11, 3215-3232.

Easterling, W.E., 1997. Why regional studies are needed in the development of full-scale integrated assessment modelling of global change processes. Global environmental change 7, 337-356.

Easton, Z.M., Fuka, D.R., White, E.D., Collick, A.S., Biruk Asharge, B., McCartney, M., Awulachew, S.B., Ahmed, A.A., Steenhuis, T.S., 2010. A multi basin SWAT model analysis of runoff and sedimentation in the Blue Nile, Ethiopia. Hydrology and Earth System Sciences Discussions 7, 3837-3878.

Eclipse, 2016. Photran - An Integrated Development Environment and Refactoring Tool for Fortran.

Edmonds, J.A., Calvin, K.V., Clarke, L.E., Janetos, A.C., Kim, S.H., Wise, M.A., McJeon, H.C., 2012. Integrated assessment modeling, Climate Change Modeling Methodology. Springer, pp. 169-209.

EEPCo, 2014. Grand Ethiopian Renaissance Dam: Significance of the Abay Basin, Addis Ababa, Ethiopia.

Ellis, E., Pontius, R., 2007. Land-use and land-cover change. Encyclopedia of earth.

Elsanabary, M.H., Gan, T.Y., 2015. Evaluation of climate anomalies impacts on the Upper Blue Nile Basin in Ethiopia using a distributed and a lumped hydrologic model. Journal of Hydrology 530, 225-240.

Elshafei, Y., Coletti, J., Sivapalan, M., Hipsey, M., 2015. A model of the socio-hydrologic dynamics in a semiarid catchment: Isolating feedbacks in the coupled human-hydrology system. Water Resources Res. 51, 6442-6471.

Ezemvelo, K., 2011. KwaZulu-Natal Land Cover 2005 V3. 1. Unpublished GIS Coverage [Clp_KZN_2005_LC_V3_1_grid_w31], Biodiversity Conservation Planning Division, Ezemvelo KZN Wildlife.

FAO, 1985. Guidelines: Land evaluation for irrigated agriculture - FAO soils bulletin 55, Rome, Italy.

FAO, 1990. Annex 2 Infiltration rate and infiltration test. FAO. [Online, Cited: November, 2016] http://www.fao.org/docrep/S8684E/s8684e0a.htm.

FAO, 2003. Country Pasture-Forage Resource Profiles-Ethiopia.

FAO, 2004a. THE GLOBAL CASSAVA DEVELOPMENT STRATEGY. Cassava for livestock feed in sub-Saharan Africa, Rome.

FAO, 2004b. Livestock Sector Brief, Ethiopia.

FAO, 2005. Livestock Sector Brief: South Africa, Livestock Information, Sector Analysis and Policy Branch, AGAL.

FAO, 2007. Geonetwork: DIGITAL SOIL MAP OF THE WORLD (DSMW), Rome, Italy.

FAO, 2013. GEONETWORK: Major soil groups of the world (FGGD). FAO, Rome, Italy.

FAO, 2014. FAO GEONETWORK: Effective Soil Depth (cm) (GeoLayer). FAO, Rome, Italy.

FAO, IIASA, ISRIC, ISSCARS, 2012. Harmonized World Soil Database (version 1.2). FAO, Rome, Italy and IIASA, Laxenburg, Austria.

FDRE, 2011. REDD: Proposal Submitted to Forest Carbon Partnership Facility. Forest Carbon Partnership Facility (FCPF).

Fernández-Giménez, M.E., Swift, D.M., 2003. Strategies for sustainable grazing management in the developing world, Proceedings of the VIIth international Rangelands congress, pp. 821-831.

Fischer, G., Nachtergaele, F., Prieler, S., Van Velthuizen, H., Verelst, L., Wiberg, D., 2008. Global agro-ecological zones assessment for agriculture (GAEZ 2008). IIASA, Laxenburg, Austria and FAO, Rome, Italy 10.

Foley, J.A., DeFries, R., Asner, G.P., Barford, C., Bonan, G., Carpenter, S.R., Chapin, F.S., Coe, M.T., Daily, G.C., Gibbs, H.K., 2005. Global consequences of land use. Science 309, 570-574.

Fürst, C., Frank, S., Witt, A., Koschke, L., Makeschin, F., 2013. Assessment of the effects of forest land use strategies on the provision of ecosystem services at regional scale. Journal of Environmental Management 127, S96-S116.

Gash, J., 1979. An analytical model of rainfall interception by forests. Quarterly Journal of the Royal Meteorological Society 105, 43-55.

Gebrehiwot, S.G., 2015. Forests, water and food security in the northwestern highlands of Ethiopia: Knowledge synthesis. Environmental Science & Policy 48, 128-136.

Gebrehiwot, S.G., Bewket, W., Gärdenäs, A.I., Bishop, K., 2014. Forest cover change over four decades in the Blue Nile Basin, Ethiopia: comparison of three watersheds. Regional Environmental Change 14, 253-266.

Geist, H.J., Lambin, E.F., 2002. Proximate Causes and Underlying Driving Forces of Tropical Deforestation: Tropical forests are disappearing as the result of many pressures, both local and regional, acting in various combinations in different geographical locations. BioScience 52, 143-150.

Ghaffari, G., Keesstra, S., Ghodousi, J., Ahmadi, H., 2010. SWAT-simulated hydrological impact of land-use change in the Zanjanrood Basin, Northwest Iran. Hydrological processes 24, 892-903.

GLC30, 2014. Globland30. National Geomatics Center of China (NGCC), Beijing, China.

Gober, P., Wheater, H.S., 2015. Debates—Perspectives on socio-hydrology: Modeling flood risk as a public policy problem. Water Resources Res. 51, 4782-4788.

Goor, Q., Halleux, C., Mohamed, Y., Tilmant, A., 2010. Optimal operation of a multipurpose multireservoir system in the Eastern Nile River Basin. Hydrology and Earth System Sciences 14, 1895-1908.

Gorelick, N., 2013. Google Earth Engine, EGU General Assembly Conference Abstracts, Vienna, p. 11997.

Gorelick, N., Hancher, M., Dixon, M., Ilyushchenko, S., Thau, D., Moore, R., 2017. Google Earth Engine: Planetary-scale geospatial analysis for everyone. Remote Sens Environ.

GRDC, 2013. River Discharge Time Series, In: GRDC (Ed.), Global Runoff Data, Koblenz, Federal Institute of Hydrology (BfG).

Gregersen, J., Gijsbers, P., Westen, S., 2007. OpenMI: Open modelling interface. Journal of Hydroinformatics 9, 175-191.

Hagen-Zanker, A., Lajoie, G., 2008. Neutral models of landscape change as benchmarks in the assessment of model performance. Landscape and Urban Planning 86, 284-296.

Halmy, M.W.A., Gessler, P.E., Hicke, J.A., Salem, B.B., 2015. Land use/land cover change detection and prediction in the north-western coastal desert of Egypt using Markov-CA. Applied Geography 63, 101-112.

Hamilton, S.H., ElSawah, S., Guillaume, J.H., Jakeman, A.J., Pierce, S.A., 2015. Integrated assessment and modelling: overview and synthesis of salient dimensions. Environmental Modelling & Software 64, 215-229.

Harbor, J.M., 1994. A practical method for estimating the impact of land-use change on surface runoff, groundwater recharge and wetland hydrology. Journal of the American Planning Association 60, 95-108.

Haregeweyn, N., Tsunekawa, A., Tsubo, M., Meshesha, D., Adgo, E., Poesen, J., Schütt, B., 2016. Analyzing the hydrologic effects of region-wide land and water development interventions: a case study of the Upper Blue Nile basin. Regional Environmental Change 16, 951-966.

Harris, G., 2002. Integrated assessment and modelling: an essential way of doing science. Environmental Modelling & Software 17, 201-207.

Hengl, T., de Jesus, J.M., MacMillan, R.A., Batjes, N.H., Heuvelink, G.B., Ribeiro, E., Samuel-Rosa, A., Kempen, B., Leenaars, J.G., Walsh, M.G., 2014. SoilGrids1km—global soil information based on automated mapping. PloS one 9, e105992.

Herold, M., Couclelis, H., Clarke, K.C., 2005. The role of spatial metrics in the analysis and modeling of urban land use change. Computers, Environment and Urban Systems 29, 369-399.

Holechek, J.L., Pieper, R.D., Herbel, C.H., 1995. Range management: principles and practices. Prentice-Hall.

Hurkens, J., Hahn, B., Van Delden, H., 2008. Using the GEONAMICA software environment for integrated dynamic spatial modelling, Proceedings of the iEMSs Fourth Biennial Meeting: Integrating Sciences and Information Technology for Environmental Assessment and Decision Making. International Environmental Modelling and Software Society, Barcelona, Spain, pp. 751-758.

Hurni, H., Tato, K., Zeleke, G., 2005. The implications of changes in population, land use, and land management for surface runoff in the upper Nile Basin Area of Ethiopia. Mountain Research and Development 25, 147-154.

Ichii, K., Wang, W., Hashimoto, H., Yang, F., Votava, P., Michaelis, A.R., Nemani, R.R., 2009. Refinement of rooting depths using satellite-based evapotranspiration seasonality for ecosystem modeling in California. Agric For Meteorol 149, 1907-1918.

ILRI, FAO, 1992. Technical paper 1: Soil classification and characterization, In: Tripathl, B.R., Psychas, P.J. (Eds.), Alley Farming Research Network for Africa. International Livestock Centre for Africa, Addis Ababa.

Jackson, R., Canadell, J., Ehleringer, J.R., Mooney, H., Sala, O., Schulze, E., 1996. A global analysis of root distributions for terrestrial biomes. Oecologia 108, 389-411.

Jakeman, A.J., Barreteau, O., Borsuk, M.E., Elsawah, S., Hamilton, S.H., Henriksen, H.J., Kuikka, S., Maier, H.R., Rizzoli, A.E., Van Delden, H., 2013. Selecting among five common modelling approaches for integrated environmental assessment and management. Environmental Modelling & Software 47, 159-181.

Jakeman, A.J., Letcher, R.A., 2003. Integrated assessment and modelling: features, principles and examples for catchment management. Environmental Modelling & Software 18, 491-501.

Jayne, T.S., Yamano, T., Weber, M.T., Tschirley, D., Benfica, R., Chapoto, A., Zulu, B., 2003. Smallholder income and land distribution in Africa: implications for poverty reduction strategies. Food policy 28, 253-275.

Johnston, R., Cox, D., Waldron, S., 2014. Assessment of Ecosystem Services in the Uthukela District Municipality.

Jolliffe, I., 2005. Principal component analysis. Wiley Online Library.

Juhos, K., Szabó, S., Ladányi, M., 2016. Explore the influence of soil quality on crop yield using statistically-derived pedological indicators. Ecological Indicators 63, 366-373.

Jun, C., Ban, Y., Li, S., 2014. China: Open access to Earth land-cover map. Nature 514, 434-434.

Kaiser, H.F., 1960. The application of electronic computers to factor analysis. Educational and psychological measurement.

Kalyanapu, A.J., Burian, S.J., McPherson, T.N., 2010. Effect of land use-based surface roughness on hydrologic model output. Journal of Spatial Hydrology 9.

Karlsson, I.B., Sonnenborg, T.O., Refsgaard, J.C., Trolle, D., Børgesen, C.D., Olesen, J.E., Jeppesen, E., Jensen, K.H., 2016. Combined effects of climate models, hydrological model structures and land use scenarios on hydrological impacts of climate change. Journal of Hydrology 535, 301-317.

Kelly, R.A., Jakeman, A.J., Barreteau, O., Borsuk, M.E., ElSawah, S., Hamilton, S.H., Henriksen, H.J., Kuikka, S., Maier, H.R., Rizzoli, A.E., 2013. Selecting among five common modelling approaches for integrated environmental assessment and management. Environ. Model. Software 47, 159-181.

Kemp, D., Michalk, D., Virgona, J., 2000. Towards more sustainable pastures: lessons learnt. Animal Production Science 40, 343-356.

Khan, D., Samadder, S.R., 2015. A simplified multi-criteria evaluation model for landfill site ranking and selection based on AHP and GIS. Journal of Environmental Engineering and Landscape Management 23, 267-278.

Koch, F.J., van Griensven, A., Uhlenbrook, S., Tekleab, S., Teferi, E., 2012. The Effects of land use change on hydrological responses in the choke mountain range (Ethiopia)-a new approach addressing land use dynamics in the model SWAT, Proceedings of 2012 international congress on environmental modeling and software managing resources of a limited planet, sixth biennial meeting, Leipzig, Germany. Citeseer, pp. 1-5.

Köhler, L., Mulligan, M., Schellekens, J., Schmid, S., Tobón, C., 2006. Final Technical Report DFID-FRP Project no. R7991 Hydrological impacts of converting tropical montane cloud forest to pasture, with initial reference to northern Costa Rica.

Koschke, L., Fuerst, C., Frank, S., Makeschin, F., 2012. A multi-criteria approach for an integrated land-cover-based assessment of ecosystem services provision to support landscape planning. Ecological Indicators 21, 54-66.

Kotch, F.J., van Griensven, A., Uhlenbrook, S., Tekleab, S., Teferi, E., 2012. The Effects of Land use Change on Hydrological Responses in the Choke Mountain Range (Ethiopia)-A new Approach Addressing Land Use Dynamics in the Model SWAT, In: Seppelt, R., Voinov, A., Lange, S., Bankamp, D. (Eds.), IEMSs2012, Leipzig, Germany.

Krysanova, V., Müller-Wohlfeil, D.-I., Becker, A., 1998. Development and test of a spatially distributed hydrological/water quality model for mesoscale watersheds. Ecological Modelling 106, 261-289.

Krysanova, V., Wechsung, F., 2000. SWIM: User Manual.

Kuhnert, M., Voinov, A., Seppelt, R., 2005. Comparing raster map comparison algorithms for spatial modeling and analysis. Photogrammetric Engineering and Remote Sensing 71, 975.

Lambin, E.F., Geist, H.J., Lepers, E., 2003. Dynamics of land-use and land-cover change in tropical regions. Annual review of environment and resources 28, 205-241.

Lambin, E.F., Meyfroidt, P., 2010. Land use transitions: Socio-ecological feedback versus socio-economic change. Land Use Policy 27, 108-118.

Lambin, E.F., Meyfroidt, P., 2011. Global land use change, economic globalization, and the looming land scarcity. Proceedings of the National Academy of Sciences 108, 3465-3472.

Lambin, E.F., Turner, B.L., Geist, H.J., Agbola, S.B., Angelsen, A., Bruce, J.W., Coomes, O.T., Dirzo, R., Fischer, G., Folke, C., 2001. The causes of land-use and land-cover change: moving beyond the myths. Global environmental change 11, 261-269.

Landis, J.R., Koch, G.G., 1977a. An application of hierarchical kappa-type statistics in the assessment of majority agreement among multiple observers. Biometrics, 363-374.

Landis, J.R., Koch, G.G., 1977b. The measurement of observer agreement for categorical data. Biometrics, 159-174.

Laniak, G.F., Olchin, G., Goodall, J., Voinov, A., Hill, M., Glynn, P., Whelan, G., Geller, G., Quinn, N., Blind, M., 2013. Integrated environmental modeling: a vision and roadmap for the future. Environmental Modelling & Software 39, 3-23.

Le Maitre, D.C., Kotzee, I.M., O'Farrell, P.J., 2014. Impacts of land-cover change on the water flow regulation ecosystem service: Invasive alien plants, fire and their policy implications. Land Use Policy 36, 171-181.

Lehohla, P., 2012. Census 2011 Municipal Report KwaZulu Natal. Statistics South Africa, Pretoria.

Leistritz, F.L., Thompson, F., Leitch, J.A., 1992. Economic impact of leafy spurge (Euphorbia esula) in North Dakota. Weed Sci, 275-280.

Lemaire, G., Hodgson, J., Chabbi, A., 2011. Grassland productivity and ecosystem services. CABI.

Letcher, R.A., Croke, B.F., Jakeman, A.J., 2007. Integrated assessment modelling for water resource allocation and management: A generalised conceptual framework. Environmental Modelling & Software 22, 733-742.

Li, J., Mao, X., Li, M., 2017. Modeling hydrological processes in oasis of Heihe River Basin by landscape unit-based conceptual models integrated with FEFLOW and GIS. Agric Water Manag 179, 338-351.

Li, K., Coe, M., Ramankutty, N., De Jong, R., 2007. Modeling the hydrological impact of land-use change in West Africa. Journal of hydrology 337, 258-268.

Liu, Y., Gupta, H., Springer, E., Wagener, T., 2008. Linking science with environmental decision making: Experiences from an integrated modeling approach to supporting sustainable water resources management. Environmental Modelling & Software 23, 846-858.

Mabbutt, J.A., 1984. A new global assessment of the status and trends of desertification. Environmental Conservation 11, 103-113.

Malczewski, J., 2004. GIS-based land-use suitability analysis: a critical overview. Progress in planning 62, 3-65.

Malczewski, J., 2006. GIS-based multicriteria decision analysis: a survey of the literature. International Journal of Geographical Information Science 20, 703-726.

Malczewski, J., Rinner, C., 2015. Multicriteria decision analysis in geographic information science. Springer.

Mango, L.M., Melesse, A.M., McClain, M.E., Gann, D., Setegn, S., 2011. Land use and climate change impacts on the hydrology of the upper Mara River Basin, Kenya: results of a modeling study to support better resource management. Hydrology and Earth System Sciences 15, 2245.

Marinoni, O., 2004. Implementation of the analytical hierarchy process with VBA in ArcGIS. Computers & Geosciences 30, 637-646.

Marinoni, O., 2009. AHP 1.1 – Decision support tool for ArcGIS.

McAfee, A., Brynjolfsson, E., 2012. Big data: the management revolution. Harvard business review, 60-66, 68, 128.

McColl, C., Aggett, G., 2007. Land-use forecasting and hydrologic model integration for improved land-use decision support. J Environ Manag 84, 494-512.

Mengistu, A., 2006. Forage resources profile of Ethiopia. FAO forage resources profiles, FAO, Rome, Italy.

Metzger, M., Rounsevell, M., Acosta-Michlik, L., Leemans, R., Schröter, D., 2006. The vulnerability of ecosystem services to land use change. Agriculture, Ecosystems & Environment 114, 69-85.

Millennium Ecosystem Assessment, 2005. Millennium ecosystem assessment. Ecosystems and Human Well-Being: Biodiversity Synthesis, Published by World Resources Institute, Washington, DC.

Mimler, M., Priess, J.A., 2008. Design and complementation of a generic modeling framework-a platform for integrated land use modeling. kassel university press GmbH.

Mishra, S.K., Singh, V., 2013. Soil conservation service curve number (SCS-CN) methodology. Springer Science & Business Media.

Monier, E., Kicklighter, D., Ejaz, Q., Winchester, N., Paltsev, S., Reilly, J., 2016. Integrated modeling of land-use change: the role of coupling, interactions and feedbacks between the human and Earth systems, AGU Fall Meeting Abstracts.

Monteith, J., Moss, C., 1977. Climate and the efficiency of crop production in Britain [and discussion]. Philosophical Transactions of the Royal Society of London. B, Biological Sciences 281, 277-294.

Moore, R., Hansen, M., 2011. Google Earth Engine: a new cloud-computing platform for global-scale earth observation data and analysis, AGU Fall Meeting Abstracts, p. 02.

Moriasi, D., Arnold, J., Van Liew, M., Bingner, R., Harmel, R., Veith, T., 2007. Model evaluation guidelines for systematic quantification of accuracy in watershed simulations. Trans. ASABE 50, 885-900.

Motuma, M., Suryabhagavan, K., Balakrishnan, M., 2016. Land suitability analysis for wheat and sorghum crops in Wogdie District, South Wollo, Ethiopia, using geospatial tools. Applied Geomatics 8, 57-66.

Mustak, S., Baghmar, N., Singh, S., 2015. Land Suitability Modeling for Gram Crop using Remote Sensing and GIS: A Case Study of Seonath Basin, India. Bulletin of Environmental and Scientific Research 4.

Narasimhan, B., Allen, P., Coffman, S., Arnold, J., Srinivasan, R., 2017. Development and Testing of a Physically Based Model of Streambank Erosion for Coupling with a

Basin-Scale Hydrologic Model SWAT. JAWRA Journal of the American Water Resources Association 53, 344-364.

Nash, J.E., Sutcliffe, J.V., 1970. River flow forecasting through conceptual models part I—A discussion of principles. Journal of Hydrology 10, 282-290.

Nativi, S., Mazzetti, P., Santoro, M., Papeschi, F., Craglia, M., Ochiai, O., 2015. Big Data challenges in building the Global Earth Observation System of Systems. Environmental Modelling & Software 68, 1-26.

Neke, K.S., Du Plessis, M.A., 2004. The threat of transformation: quantifying the vulnerability of grasslands in South Africa. Conservation Biology 18, 466-477.

Nie, W., Yuan, Y., Kepner, W., Nash, M.S., Jackson, M., Erickson, C., 2011. Assessing impacts of Landuse and Landcover changes on hydrology for the upper San Pedro watershed. Journal of Hydrology 407, 105-114.

Niehoff, D., Fritsch, U., Bronstert, A., 2002. Land-use impacts on storm-runoff generation: scenarios of land-use change and simulation of hydrological response in a meso-scale catchment in SW-Germany. Journal of Hydrology 267, 80-93.

Norgaard, R.B., 2010. Ecosystem services: From eye-opening metaphor to complexity blinder. Ecological Economics 69, 1219-1227.

O'Connell, P., Ewen, J., O'Donnell, G., Quinn, P., 2007. Is there a link between agricultural land-use management and flooding? Hydrology and Earth System Sciences 11, 96-107.

Olaniyi, A., Ajiboye, A., Abdullah, A., Ramli, M., Sood, A., 2015. Agricultural land use suitability assessment in Malaysia. Bulgarian Journal of Agricultural Science 21, 560-572.

Olmedo, M.T.C., Pontius, R.G., Paegelow, M., Mas, J.-F., 2015. Comparison of simulation models in terms of quantity and allocation of land change. Environmental Modelling & Software 69, 214-221.

Olson, D.M., Dinerstein, E., 1998. The Global 200: a representation approach to conserving the Earth's most biologically valuable ecoregions. Conservation Biology 12, 502-515.

Ott, B., Uhlenbrook, S., 2004. Quantifying the impact of land-use changes at the event and seasonal time scale using a process-oriented catchment model. Hydrology and Earth System Sciences 8, 62-78.

Ozturk, D., Batuk, F., 2011. Implementation of GIS-based multicriteria decision analysis with VB in ArcGIS. International Journal of Information Technology & Decision Making 10, 1023-1042.

Pankhurst, A., 2010. LAND DEGRADATION. Water Resources Management in Ethiopia: Implications for the Nile Basin, 213.

Peng, D., Lee, F.C., Boroyevich, D., 2002. A novel SVM algorithm for multilevel three-phase converters, Power Electronics Specialists Conference, 2002. pesc 02. 2002 IEEE 33rd Annual. IEEE, pp. 509-513.

Pontius Jr, R.G., Boersma, W., Castella, J.-C., Clarke, K., de Nijs, T., Dietzel, C., Duan, Z., Fotsing, E., Goldstein, N., Kok, K., 2008. Comparing the input, output, and validation maps for several models of land change. The Annals of Regional Science 42, 11-37.

Pontius Jr, R.G., Millones, M., 2011. Death to Kappa: birth of quantity disagreement and allocation disagreement for accuracy assessment. International Journal of Remote Sensing 32, 4407-4429.

Pontius, R.G., 2000. Quantification error versus location error in comparison of categorical maps. Photogrammetric Engineering and Remote Sensing 66, 1011-1016.

Pontius, R.G., 2004. Useful techniques of validation for spatially explicit land-change models. Ecological Modelling 179, 445-461.

Pontius, R.G., Millones, M., 2011. Death to Kappa: birth of quantity disagreement and allocation disagreement for accuracy assessment. International Journal of Remote Sensing 32, 4407-4429.

Prakash, T., 2003. Land suitability analysis for agricultural crops: A fuzzy Multicriteria Decision Making Approach. MS Theses international institute for geo-information science and earth observation enschede, the netherland.

Pramanik, M.K., 2016. Site suitability analysis for agricultural land use of Darjeeling district using AHP and GIS techniques. Modeling Earth Systems and Environment 2, 1-22.

Priess, J., Mimler, M., Klein, A.-M., Schwarze, S., Tscharntke, T., Steffan-Dewenter, I., 2007. Linking deforestation scenarios to pollination services and economic returns in coffee agroforestry systems. Ecological Applications 17, 407-417.

Priess, J.A., Schweitzer, C., Wimmer, F., Batkhishig, O., Mimler, M., 2011. The consequences of land-use change and water demands in Central Mongolia. Land Use Policy 28, 4-10.

Rendana, M., Rahim, S.A., Idris, W.M.R., Lihan, T., Rahman, Z.A., 2015. CA-Markov for Predicting Land Use Changes in Tropical Catchment Area: A Case Study in Cameron Highland, Malaysia. Journal of Applied Sciences 15, 689.

Reshmidevi, T., Eldho, T., Jana, R., 2009. A GIS-integrated fuzzy rule-based inference system for land suitability evaluation in agricultural watersheds. Agricultural Systems 101, 101-109.

Reuters, T., 2012. Web of Science.

Reyers, B., Fairbanks, D., Van Jaarsveld, A., Thompson, M., 2001. Priority areas for the conservation of South African vegetation: a coarse-filter approach. Diversity and Distributions 7, 79-95.

Reyers, B., O'Farrell, P.J., Cowling, R.M., Egoh, B.N., Le Maitre, D.C., Vlok, J.H., 2009. Ecosystem services, land-cover change, and stakeholders: finding a sustainable foothold for a semiarid biodiversity hotspot.

Rientjes, T., Haile, A., Kebede, E., Mannaerts, C., Habib, E., Steenhuis, T., 2011. Changes in land cover, rainfall and stream flow in Upper Gilgel Abbay catchment, Blue Nile basin–Ethiopia. Hydrol. Earth Syst. Sci 15, 1979-1989.

Rindfuss, R.R., Entwisle, B., Walsh, S.J., An, L., Badenoch, N., Brown, D.G., Deadman, P., Evans, T.P., Fox, J., Geoghegan, J., 2008. Land use change: complexity and comparisons. Journal of Land Use Science 3, 1-10.

Robinson, D.T., Di Vittorio, A., Alexander, P., Arneth, A., Barton, C.M., Brown, D.G., Kettner, A., Lemmen, C., O'Neill, B.C., Janssen, M., Pugh, T.A.M., Rabin, S.S., Rounsevell, M., Syvitski, J.P., Ullah, I., Verburg, P.H., 2017. Modelling feedbacks between human and natural processes in the land system. Earth Syst. Dynam. Discuss. 2017, 1-47.

Rossiter, D.G., 1996. A theoretical framework for land evaluation. Geoderma 72, 165-190.

Rotmans, J., van Asselt, M., 1999. Integrated assessment modelling, Climate change: An integrated perspective. Springer, pp. 239-275.

RSA, 1998. National Environmental Management Act (Act No. 107 of 1998). Government Gazette, South Africa 401.

Rudel, T.K., Schneider, L., Uriarte, M., Turner, B.L., DeFries, R., Lawrence, D., Geoghegan, J., Hecht, S., Ickowitz, A., Lambin, E.F., 2009. Agricultural intensification and changes in cultivated areas, 1970–2005. Proceedings of the National Academy of Sciences 106, 20675-20680.

Saaty, R.W., 1987. The analytic hierarchy process—what it is and how it is used. Mathematical modelling 9, 161-176.

Saaty, T.L., 1980. The analytic hierarchy process: planning, priority setting, resource allocation. McGraw-Hill International Book Co.

Saaty, T.L., 1988. What is the analytic hierarchy process? Springer.

Saaty, T.L., 2008. Decision making with the analytic hierarchy process. International journal of services sciences 1, 83-98.

Sajikumar, N., Remya, R., 2015. Impact of land cover and land use change on runoff characteristics. J Environ Manag 161, 460-468.

Salomon, M., 2006. THE PEOPLE AND THEIR CATTLE COMMUNAL GRAZING IN THE NORTHERN DRAKENSBERG, Proceedings of the 40th Annual Conference of the South African Society for Agricultural Extension. Berg en Dal, pp. 9-11.

Scarnecchia, D.L., 1985. The animal-unit and animal-unit-equivalent concepts in range science. J Range Manag, 346-349.

Schellekens, J., 2014. wflow Documentation. Deltares. [Online, Cited: Novermber 20, 2016]. https://publicwiki.deltares.nl/display/OpenS/Home. Deltares. Available: https://publicwiki.deltares.nl/display/OpenS/Home, Accessed: Novermber 20, 2016.

Schenk, H., Jackson, R., 2009. Islscp ii ecosystem rooting depths. ISLSCP Initiative II Collection. Data set. Available: http://daac. ornl. gov/, from Oak Ridge National Laboratory Distributed Active Archive Center, Oak Ridge, Tennessee, USA doi 10.

Scherr, S.J., McNeely, J.A., 2008. Biodiversity conservation and agricultural sustainability: towards a new paradigm of 'ecoagriculture'landscapes. Philosophical Transactions of the Royal Society of London B: Biological Sciences 363, 477-494.

Schweitzer, C., Priess, J.A., Das, S., 2011. A generic framework for land-use modelling. Environmental Modelling & Software 26, 1052-1055

Scott-Shaw, B., Schulze, R., 2013. Use of a Grassland Biomass Model for Applications in Management Studies of Tall and Short Grassveld.

Scott, D.F., Lesch, W., 1997. Streamflow responses to afforestation with Eucalyptus grandis and Pinus patula and to felling in the Mokobulaan experimental catchments, South Africa. Journal of Hydrology 199, 360-377.

Serneels, S., Lambin, E.F., 2001. Proximate causes of land-use change in Narok District, Kenya: a spatial statistical model. Agriculture, Ecosystems & Environment 85, 65-81.

Serpa, D., Nunes, J., Santos, J., Sampaio, E., Jacinto, R., Veiga, S., Lima, J., Moreira, M., Corte-Real, J., Keizer, J., 2015. Impacts of climate and land use changes on the hydrological and erosion processes of two contrasting Mediterranean catchments. Sci Total Environ 538, 64-77.

Setegn, S.G., Srinivasan, R., Dargahi, B., 2008. Hydrological modelling in the Lake Tana Basin, Ethiopia using SWAT model. Open Hydrology Journal 2, 49-62.

Setegn, S.G., Srinivasan, R., Dargahi, B., Melesse, A.M., 2009. Spatial delineation of soil erosion vulnerability in the Lake Tana Basin, Ethiopia. Hydrological Processes 23, 3738-3750.

Shalaby, A., Ouma, Y., Tateishi, R., 2006. Land suitability assessment for perennial crops using remote sensing and Geographic Information Systems: A case study in northwestern Egypt: (Bewertung der Eignung von Standorten zum Anbau von mehrjährigen Fruchtarten mittels Fernerkundung und GIS: eine Fallstudie in Nord-West Ägypten). Archives of Agronomy and Soil Science 52, 243-261.

Sheng, T., 1990. Watershed Management Field Manual. FAO conservation guide 13.

Sieber, R., 2006. Public participation geographic information systems: A literature review and framework. Annals of the Association of American Geographers 96, 491-507.

Silva, S., Alçada-Almeida, L., Dias, L.C., 2014. Development of a Web-based Multi-criteria Spatial Decision Support System for the assessment of environmental sustainability of dairy farms. Computers and Electronics in Agriculture 108, 46-57.

Simane, B., Zaitchik, B.F., Ozdogan, M., 2013. Agroecosystem analysis of the Choke Mountain watersheds, Ethiopia. Sustainability 5, 592-616.

Siriwardena, L., Finlayson, B.L., McMahon, T.A., 2006. The impact of land use change on catchment hydrology in large catchments: The Comet River, Central Queensland, Australia. Journal of Hydrology 326, 199-214.

Sivapalan, M., Konar, M., Srinivasan, V., Chhatre, A., Wutich, A., Scott, C., Wescoat, J., Rodríguez-Iturbe, I., 2014. Socio-hydrology: Use-inspired water sustainability science for the Anthropocene. Earth's Future 2, 225-230.

Sivapalan, M., Savenije, H.H., Blöschl, G., 2012. Socio-hydrology: A new science of people and water. Hydrological Processes 26, 1270-1276.

Skånes, H., Bunce, R., 1997. Directions of landscape change (1741–1993) in Virestad, Sweden—characterised by multivariate analysis. Landscape and Urban Planning 38, 61-75.

Smit, H., Muche, R., Ahlers, R., van der Zaag, P., 2017. The Political Morphology of Drainage—How Gully Formation Links to State Formation in the Choke Mountains of Ethiopia. World Development 98, 231-244.

South African Environmental Affairs, 1990. National Land Cover Data for SA, South Africa.

Spehn, E.M., Liberman, M., Korner, C., 2006. Land use change and mountain biodiversity. CRC Press.

Steenhuis, T.S., Collick, A.S., Easton, Z.M., Leggesse, E.S., Bayabil, H.K., White, E.D., Awulachew, S.B., Adgo, E., Ahmed, A.A., 2009. Predicting discharge and sediment for the Abay (Blue Nile) with a simple model. Hydrological Processes, n/a-n/a.

Steiner, F., McSherry, L., Cohen, J., 2000. Land suitability analysis for the upper Gila River watershed. Landscape and Urban Planning 50, 199-214.

Tainton, N., 1999 Rangeland Carrying Capacity. (ed)KwaZulu-Natal Department of Agriculture and Environmental Affairs 1999. Veld

in KwaZulu-Natal, South Africa.

Tainton, N., Edwards, P., Mentis, M., 1980. A revised method for assessing veld condition. Proceedings of the Annual Congresses of the Grassland Society of Southern Africa 15, 37-42.

Tang, Q., 2016. Terrestrial Water Cycle and Climate Change: Natural and Human-Induced Impacts. John Wiley & Sons.

Teferi, E., Bewket, W., Uhlenbrook, S., Wenninger, J., 2013a. Understanding recent land use and land cover dynamics in the source region of the Upper Blue Nile, Ethiopia: spatially explicit statistical modeling of systematic transitions. Agriculture, Ecosystems & Environment 165, 98-117.

Teferi, E., Bewket, W., Uhlenbrook, S., Wenninger, J., 2013b. Understanding recent land use/ cover dynamics in the source region of the Upper Blue Nile, Ethiopia: systematic and random transitions. Agriculture, Ecosystems & Environment 165, 98-117.

Tekleab, S., Mohamed, Y., Uhlenbrook, S., Wenninger, J., 2014a. Hydrologic responses to land cover change: the case of Jedeb mesoscale catchment, Abay/Upper Blue Nile basin, Ethiopia. Hydrological Processes 28, 5149-5161.

Tekleab, S., Wenninger, J., Uhlenbrook, S., 2014b. Characterisation of stable isotopes to identify residence times and runoff components in two meso-scale catchments in the Abay/Upper Blue Nile basin, Ethiopia. Hydrology and Earth System Sciences 18, 2415-2431.

Terribile, F., Agrillo, A., Bonfante, A., Buscemi, G., Colandrea, M., D'Antonio, A., De Mascellis, R., De Michele, C., Langella, G., Manna, P., 2015. A Web-based spatial decision supporting system for land management and soil conservation. Solid Earth 6, 903.

Tesfaye, A., Brouwer, R., 2012. Testing participation constraints in contract design for sustainable soil conservation in Ethiopia. Ecological Economics 73, 168-178.

Tong, S.T., Chen, W., 2002. Modeling the relationship between land use and surface water quality. Journal of Environmental Management 66, 377-393.

Trabucco, A., Zomer, R., 2010. Global soil water balance geospatial database. CGIAR Consortium for Spatial Information. Published online, available from the CGIARCSI GeoPortal at http://cgiar-csi. org.

Tscharntke, T., Clough, Y., Wanger, T.C., Jackson, L., Motzke, I., Perfecto, I., Vandermeer, J., Whitbread, A., 2012. Global food security, biodiversity conservation and the future of agricultural intensification. Biological conservation 151, 53-59.

Turner, B., Skole, D., Sanderson, S., Fischer, G., Fresco, L., Leemans, R., 1995. Land-use and land-cover change, International Geosphere-Biosphere Programme, Stockholm; Report, 35.

UNEP-WCMC, 2012. The World Database on Protected Areas (WDPA), Cambridge, UK.

Urgesa, A.A., Abegaz, A., Bahir, A.L., Antille, D.L., 2016. Population growth and other factors affecting land-use and land-cover changes in north-eastern Wollega, Ethiopia. Trop. Agric 41, 040298-040214.

USDA, 2016. Natural Resources Conservation Service: Part 618 – Soil Properties and Qualities. United States Department of Agriculture. [Online] [Cited: November, 2016] http://soils.usda.gov/technical/handbook/contents/part618ex.html.

Van Deursen, W., 1995. Geographical information systems and dynamic models.

Van Genuchten, M.T., 1980. A closed-form equation for predicting the hydraulic conductivity of unsaturated soils. Soil Sci Soc Am J 44, 892-898.

van Griensven, A., Ndomba, P., Yalew, S., Kilonzo, F., 2012. Critical review of SWAT applications in the upper Nile basin countries. Hydrol. Earth Syst. Sci. 16, 3371-3381.

van Oel, P.R., Krol, M.S., Hoekstra, A.Y., Taddei, R.R., 2010. Feedback mechanisms between water availability and water use in a semi-arid river basin: A spatially explicit multi-agent simulation approach. Environmental Modelling & Software 25, 433-443.

Van Orshoven, J., Terres, J., Toth, T., 2012. Updated Common bio-physical criteria to define natural constraints for agriculture in Europe. Ispra. Internet: http://agrienv. jrc. ec. europa. eu. DOI: http://dx. doi. org/10.2788/91182.

van Vliet, J., Bregt, A.K., Hagen-Zanker, A., 2011. Revisiting Kappa to account for change in the accuracy assessment of land-use change models. Ecological Modelling 222, 1367-1375.

van Vliet, J., Naus, N., van Lammeren, R.J., Bregt, A.K., Hurkens, J., van Delden, H., 2013. Measuring the neighbourhood effect to calibrate land use models. Computers, Environment and Urban Systems 41, 55-64.

Veldkamp, A., Lambin, E., 2001. Predicting land-use change. Agriculture, Ecosystems & Environment 85, 1-6.

Verburg, P.H., Dearing, J.A., Dyke, J.G., van der Leeuw, S., Seitzinger, S., Steffen, W., Syvitski, J., 2016. Methods and approaches to modelling the Anthropocene. Global Environ. Change 39, 328-340.

Verburg, P.H., Schot, P.P., Dijst, M.J., Veldkamp, A., 2004. Land use change modelling: current practice and research priorities. GeoJournal 61, 309-324.

Vertessy, R.A., Elsenbeer, H., 1999. Distributed modeling of storm flow generation in an Amazonian rain forest catchment: Effects of model parameterization. Water Resources Research 35, 2173-2187.

Visser, H., De Nijs, T., 2006. The map comparison kit. Environmental Modelling & Software 21, 346-358.

Vitolo, C., Elkhatib, Y., Reusser, D., Macleod, C.J., Buytaert, W., 2015. Web technologies for environmental Big Data. Environmental Modelling & Software 63, 185-198.

Voinov, A., Shugart, H.H., 2013. 'Integronsters', integral and integrated modeling. Environ. Model. Software 39, 149-158.

Wagner, P.D., Bhallamudi, S.M., Narasimhan, B., Kantakumar, L.N., Sudheer, K., Kumar, S., Schneider, K., Fiener, P., 2016. Dynamic integration of land use changes in a hydrologic assessment of a rapidly developing Indian catchment. Sci Total Environ 539, 153-164.

Wagner, P.D., Bhallamudi, S.M., Narasimhan, B., Kumar, S., Fohrer, N., Fiener, P., 2017. Comparing the effects of dynamic versus static representations of land use change in hydrologic impact assessments. Environ. Model. Software.

Wale, A., Collick, A.S., Rossiter, D.G., Langan, S., Steenhuis, T.S., 2013. Realistic assessment of irrigation potential in the Lake Tana basin, Ethiopia.

Wall, M., 1996. GAlib: A C++ library of genetic algorithm components. Mechanical Engineering Department, Massachusetts Institute of Technology 87, 54.

Wallace, K.J., 2007. Classification of ecosystem services: problems and solutions. Biological conservation 139, 235-246.

Wang, F., 1994. The use of artificial neural networks in a geographical information system for agricultural land-suitability assessment. Environment and planning A 26, 265-284.

Warburton, M.L., Schulze, R.E., Jewitt, G.P., 2012. Hydrological impacts of land use change in three diverse South African catchments. Journal of Hydrology 414, 118-135.

Watkinson, A., Ormerod, S., 2001. Grasslands, grazing and biodiversity: editors' introduction. Journal of Applied Ecology 38, 233-237.

Weedon, G., Gomes, S., Viterbo, P., Österle, H., Adam, J., Bellouin, N., Boucher, O., Best, M., 2010. The WATCH forcing data 1958–2001: A meteorological forcing dataset for land surface and hydrological models. WATCH, editor. WATCH Technical Report 22.

Wegener, M., 2004. Overview of land-use transport models. Handbook of transport geography and spatial systems 5, 127-146.

Wheater, H., Gober, P., 2011. Towards a new paradigm of Socio-Hydrology; insights from the Saskatchewan River Basin, AGU Fall Meeting Abstracts, p. 1134.

White, E.D., Easton, Z.M., Fuka, D.R., Collick, A.S., Adgo, E., McCartney, M., Awulachew, S.B., Selassie, Y.G., Steenhuis, T.S., 2011. Development and application of a physically based landscape water balance in the SWAT model. Hydrological Processes 25, 915-925.

White, R.P., Murray, S., Rohweder, M., Prince, S.D., Thompson, K.M., 2000. Grassland ecosystems. World Resources Institute Washington, DC, USA.

Woldegiorgis, B., 2013. Impacts of Landuse Change on Hydrological Extremes in Ribb-Gumara Catchments. MSc Thesis. Katholieke Universiteit Leuven.

Woldesenbet, T.A., Elagib, N.A., Ribbe, L., Heinrich, J., 2017. Hydrological responses to land use/cover changes in the source region of the Upper Blue Nile Basin, Ethiopia. Sci Total Environ 575, 724-741.

Yalew, S., Pilz, T., Schweitzer, C., Liersch, S., van der Kwast, J., Mul, M.L., van Griensven, A., van der Zaag, P., 2014. Dynamic Feedback between Land Use and Hydrology for Ecosystem Services Assessment.

Yalew, S., Teferi, E., van Griensven, A., Uhlenbrook, S., Mul, M., van der Kwast, J., van der Zaag, P., 2012. Land Use Change and Suitability Assessment in the Upper Blue Nile Basin Under Water Resources and Socio-economic Constraints: A Drive Towards a Decision Support System.

Yalew, S., van Griensven, A., van der Zaag, P., 2016a. AgriSuit: A web-based GIS-MCDA framework for agricultural land suitability assessment. Computers and Electronics in Agriculture 128, 1-8.

Yalew, S.G., Mul, M.L., van Griensven, A., Teferi, E., Priess, J., Schweitzer, C., van Der Zaag, P., 2016b. Land-Use Change Modelling in the Upper Blue Nile Basin. Environments 3, 21.

Yalew, S.G., van Griensven, A., Mul, M.L., van der Zaag, P., 2016c. Land suitability analysis for agriculture in the Abbay basin using remote sensing, GIS and AHP techniques. Modeling Earth Systems and Environment 2, 1-14.

Yan, B., Fang, N., Zhang, P., Shi, Z., 2013. Impacts of land use change on watershed streamflow and sediment yield: an assessment using hydrologic modelling and partial least squares regression. Journal of Hydrology 484, 26-37.

Zabihi, H., Ahmad, A., Vogeler, I., Said, M.N., Golmohammadi, M., Golein, B., Nilashi, M., 2015. Land suitability procedure for sustainable citrus planning using the application of the analytical network process approach and GIS. Computers and Electronics in Agriculture 117, 114-126.

Zeleke, G., Hurni, H., 2001. Implications of land use and land cover dynamics for mountain resource degradation in the Northwestern Ethiopian highlands. Mountain Research and Development 21, 184-191.

Zhang, J., Su, Y., Wu, J., Liang, H., 2015. GIS based land suitability assessment for tobacco production using AHP and fuzzy set in Shandong province of China. Computers and Electronics in Agriculture 114, 202-211.

Zhou, G., Wei, X., Chen, X., Zhou, P., Liu, X., Xiao, Y., Sun, G., Scott, D.F., Zhou, S., Han, L., 2015. Global pattern for the effect of climate and land cover on water yield. Nature communications 6.

Zolekar, R.B., Bhagat, V.S., 2015. Multi-criteria land suitability analysis for agriculture in hilly zone: Remote sensing and GIS approach. Computers and Electronics in Agriculture 118, 300-321.

APPENDICES

Appendix 1. WFlow model parameters

Appendix 1.1. Parameter settings for WFLOW interception module

Parameter	Description	Assigned value	Normal value range
CanopyGapFraction	Gash interception model parameter: the free throughfall coefficient	0.05	Between 0 and 1 (Asres and Awulachew, 2010; Schellekens, 2014)
EoverR	Gash interception model parameter: Ratio of average wet canopy evaporation rate over average precipitation rate	0.1	Between 0 and 1 (Schellekens, 2014)
MaxCanopyStorage	Canopy storage: used in the Gash interception model	0.5 mm	Between 0-10 mm (Asres and Awulachew, 2010; Bashar and Zaki, 2005; Breuer et al., 2003)

Appendix 1.2. WFLOW parameter settings for land-use and soil combinations

WFLOW land-use and soil parameters are combined in the model representation. In the following pages, combined parameter values for the Jedeb model setup are shown in tables. Note that the values in the 'LU-Soil Pair' column in each table represent **land use, sub-catchment, soil,** and **parameter values,** respectively. Also note that the land-use representations are: 1.Woody vegetation, 2. Plantation forest, 3.Cultivation land, 4. Grassland, 5.Bare land, 6.Wetland; and the soil representations are: 1.Eutric Leptisols, 2. Haplic Luvisols, and 3. Haplic Alisols. The notation '[0,>' means all values greater than 0.

Table A-1. First zone capacity

Parameter	LU-Soil Pair		FirstZoneCapacity	Remark
FirstZoneCapacity(mm)	[0,>	1	400	Maximum capacity of the saturated store; values between 0-20000 mm (Elsanabary and Gan, 2015; Schellekens, 2014)
	[0,>	1	7100	
	[0,>	1	7100	
	[0,>	1	7100	
	[0,>	1	7100	
	[0,>	1	7100	
	[0,>	2	3200	
	[0,>	2	13000	
	[0,>	2	10000	
	[0,>	2	12000	
	[0,>	2	10000	
	[0,>	2	10000	
	[0,>	3	400	
	[0,>	3	1200	
	[0,>	3	1000	
	[0,>	3	1100	
	[0,>	3	1000	
	[0,>	3	1000	

Table A-2. First zone saturated conductivity

Parameter	LU-Soil Pair			Remark
FirstZoneKsatVer(Ksat) (mm/day)	[0,>	1	8000	Saturated conductivity of the store at the surface. The M parameter determines how this decreases with depth ; between 0-10000 (Easton et al., 2010; USDA, 2016; van Griensven et al., 2012)
	[0,>	2	200	
	[0,>	3	100	

Table A-3. First zone minimum capacity

Parameter	LU-Soil Pair		Remarks
FirstZoneMinCapacity (mm)	1 [0,>	1	95	Minimum capacity of the saturated store; between 0-400 (Decagon Devices, 2016; van Griensven et al., 2012)
	2 [0,>	1	350	
	3 [0,>	1	350	
	4 [0,>	1	350	
	5 [0,>	1	190	
	6 [0,>	1	350	
	1 [0,>	2	95	
	2 [0,>	2	250	
	3 [0,>	2	250	
	4 [0,>	2	250	
	5 [0,>	2	90	
	6 [0,>	2	250	
	1 [0,>	3	95	
	2 [0,>	3	150	
	3 [0,>	3	150	
	4 [0,>	3	150	
	5 [0,>	3	100	
	6 [0,>	3	150	

Table A-4. Infiltration capacity of the compacted soil

Parameter	LU-Soil Pair		Remarks
InfiltCapPath	1 [0,>	1 5	Infiltration capacity of the compacted soil (or paved area) fraction
(mm/day)	2 [0,>	1 5	of each grid-cell (White et al., 2011)
	3 [0,>	1 5	
	4 [0,>	1 5	
	5 [0,>	1 3	
	6 [0,>	1 5	
	1 [0,>	2 5	
	2 [0,>	2 5	
	3 [0,>	2 5	
	4 [0,>	2 5	
	5 [0,>	2 4	
	6 [0,>	2 5	
	1 [0,>	3 3	
	2 [0,>	3 4	
	3 [0,>	3 4	
	4 [0,>	3 4	
	5 [0,>	3 3	
	6 [0,>	3 4	

Tabel A-5. Infiltration capacity of non-compacted soil

Parameter	LU-Soil Pair	Remarks
InfiltCapSoil (mm/day)	1 [0,> 1 5000	Infiltration capacity of the non-compacted soil fraction (unpaved area) of each grid cell; between 25-750[Note: Sand: < 30 mm/hour; Sandy loam: 20–30; mm/hour; Loam: 10–20 mm/hour; Clay loam: 5–10 mm/hour; Clay: 1-5 mm/hour (Dingman, 2002; FAO, 1990)]
	2 [0,> 1 5000	
	3 [0,> 1 5000	
	4 [0,> 1 5000	
	5 [0,> 1 3000	
	6 [0,> 1 4000	
	1 [0,> 2 4000	
	2 [0,> 2 4000	
	3 [0,> 2 4000	
	4 [0,> 2 4000	
	5 [0,> 2 3000	
	6 [0,> 2 4000	
	1 [0,> 3 3000	
	2 [0,> 3 4000	
	3 [0,> 3 3000	
	4 [0,> 3 3000	
	5 [0,> 3 3000	
	6 [0,> 3 4000	

Tabel A-6. Decrease of saturated conductivity with soil depth

Parameter	LU-Soil Pair		Remarks
M (mm)	[0,>	[0,>	1 12880	Soil parameter determining the decrease of saturated conductivity with depth (soil depth in mm); between 20 and 2000 (Bitew and Gebremichael, 2011; Schellekens, 2014; Van Genuchten, 1980)
	[0,>	[0,>	2 1504	
	[0,>	[0,>	3 193	

Tabel A-7. Manning's roughness

Parameter	LU-Soil Pair			Remarks
N	[0,>	[0,>	1 0.072	Manning's roughness parameter ; between 0.01- 0.4
	[0,>	[0,>	2 0.048	(Betrie et al., 2011; Kalyanapu et al., 2010; Koch et al.,
	[0,>	[0,>	3 0.036	2012)

Table A-8. Manning's N for the river

Parameter	LU-Soil Pair			Remarks
N_river	[0,>	[0,>	1 0.14	Manning's parameter for the river ; between 0.01-0.15 (Schellekens, 2014) [Mountain rivers: >0.045; Winding natural streams with weeds: 0.035; Natural streams with little vegetation: 0.025]
	[0,>	[0,>	2 0.1	
	[0,>	[0,>	3 0.027	

Table A-9. Fraction of compacted area per grid-cell

Parameter	LU-Soil Pair		Remarks
PathFrac	[0,>	[0,> 1 0.1	Fraction of compacted area per grid cell; between 0 and 1
	[0,>	[0,> 2 0.12	(Amézquita et al., 1998; Schellekens, 2014)
	[0,>	[0,> 3 0.15	

Tabel A-10. Rooting depth

RootingDepth (mm)			
1	[0,>	1 2	
2	[0,>	1 84	
3	[0,>	1 245	
4	[0,>	1 140	
5	[0,>	1 84	
6	[0,>	1 2	
1	[0,>	2 2	
2	[0,>	2 84	
3	[0,>	2 245	
4	[0,>	2 140	
5	[0,>	2 84	
6	[0,>	2 2	
1	[0,>	3 2	
2	[0,>	3 48	
3	[0,>	3 140	
4	[0,>	3 80	
5	[0,>	3 48	
6	[0,>	3 2	

Rooting depth of the vegetation ; between 0-7000

[Trees:<=7000; Shrubs : < 5100; Herbaceous: <2600

(Ichii et al., 2009; Jackson et al., 1996)]

Tabel A-11. Residual water content

Parameter	LU-Soil Pair		Remarks
thetaR	[0,> [0,>	[0,> 0.003	Residual water content; between 0.001–0.1 (Jackson et al., 1996)

Table A-12. Water content at saturation

Parameter	LU-Soil Pair			Remarks
thetaS	[0,>	[0,>	[0,> 0.3	Water content at saturation; between 0.2–0.5 (Jackson et al., 1996)

Appendix 2. Scenario simulation

Extending section 6.4, further simulation of feedback land-use change and hydrologic models with respect to total grassland biomass and grazing ecosystem services from grassland until 2030 were simulated in the coupled model. Model results were compared with those from the uncoupled model. For this simulation, both climate and land-use demand scenarios were incorporated.

In the hydrologic model, a climate projection scenario until the year 2030 was incorporated as indirect input through its effects on production of biomass in the SWIM model. Anthropogenic Greenhouse Gas (GHG) emissions are believed to be driven by population size, economic activity, lifestyle, energy use, land-use patterns, technology and climate policy (Bernstein et al., 2007). Representative Concentration Pathways (RCPs) describe four different 21st century pathways of climate change scenarios based on these climate forcing variables (Bernstein et al., 2007). RCPs include a stringent mitigation scenario (RCP2.6), two intermediate scenarios (RCP4.5 and RCP6.0) and one scenario with continuing GHG emissions (RCP8.5). Scenarios without additional efforts to constrain emissions lead to pathways ranging between RCP6.0 and RCP8.5 and RCP2.6 is representative of a scenario that aims to keep global warming likely below 2°C above pre-industrial temperatures (Bernstein et al., 2007). The RCP2.6 climate scenario was used for simulating the hydrology of the catchment until 2030. This pathway assumes that emissions will continue to rise until 2040 (IPCC, 2007) before they start declining afterwards.

In relation to scenario development for the land-use model, past trends of population, livestock, settlement, and land-use change are used. The population growth rate in the Thukela district is 0.17%/year (Lehohla, 2012). Increased livestock population in the rural community in this district, besides its commercial value, is a sign of more wealth and respect (Johnston et al., 2014; Salomon, 2006). The annual livestock growth rate for South Africa between 1990 and 2000 was reported to be 0.2%/year (FAO, 2004, 2005). For the land-use change model development in this study, we assumed the same annual livestock growth rate to continue to 2030. A business-as-usual (BAU) scenario of land-use change assuming annual growth rates of 0.17% and 0.2% for population and livestock, respectively, and associated demands for various land uses is, therefore, simulated until 2030.

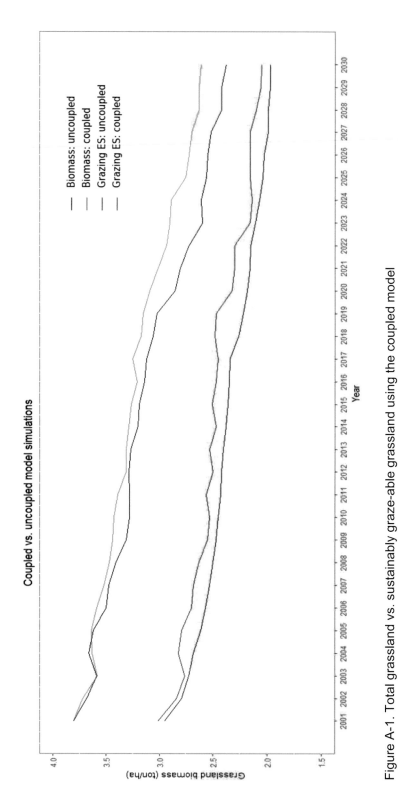

Figure A-1. Total grassland vs. sustainably graze-able grassland using the coupled model

Appendix 3. SWIM-SITE Coupling: code snippets

Appendix 3.1. SWIM output to SITE input converter

```
"""
Script to load SWIM GIS output and create ascii file for input in SITE.
"""
import os, sys
workingDir=os.getcwd()
print "Starting Script to convert SWIM GIS output into ascii raster file for SITE..."
print ""
print "Reading settings..."
print ""
### SETTINGS ###
# INPUT #
cellDir=workingDir+"\grassdata\South_Africa\PERMANENT\cell"
os.chdir(cellDir)
#GRASS raster -> already in GRASS location
mask_site="mask_site"
#mask_site = "mask_site" # mask with extent used in SITE
hyd_recl = "hyd_recl" # reclassified hydrotope raster created by ascii2swim.py to convert SWIM GIS
output into raster
#os.chdir(workingDir)
# SWIM GIS files to be converted into ascii raster file
gis_dir = workingDir+"\data\GIS"
gis_files = ("hydbio-gis.out","pre-gis.out") # may be appended (processed in a loop) # ("hydbio-
gis.out", "petmean-gis.out") # may be appended (processed in a loop)
# OUTPUT #
# location and name of ascii raster file
out_dir = workingDir+"\dataExchange"
#print "Working Dir="+workingDir      #test...........
ascii_out=("hydbio-gis.asc", "pre-gis.asc") # order corresponding to gis_files
# GRASS SETTINGS #
# initialize grass session
gisbase = os.environ['GISBASE']
gisdb = workingDir+"\grassdata"
print "gisdb="+ gisdb
```

```
#print "gis_dir="+gis_dir
location = "South_Africa"
mapset = "PERMANENT"
import grass.script as grass
import grass.script.setup as gsetup
gsetup.init(gisbase, gisdb, location, mapset)
### CALCULATION ###
print "Reclassifying SWIM GIS output file and loading into GRASS location..."
print ""
# reclassify swim gis file
if len(gis_files) != len(ascii_out):
        print "ERROR: gis_file and ascii_out don't have the same length!"
        sys.exit()
for file in gis_files:
grass.run_command("r.reclass", input=hyd_recl, output=file+"_t", rules=gis_dir+"\\"+file,
overwrite=True)
# mask catchment to smaller extend modelled by SITE
grass.run_command("r.mask", "r")
grass.run_command("r.mask", input=mask_site)
print "Writing into ascii file..."
print ""
# write output ascii file
for i in range(len(ascii_out)):
grass.run_command("r.out.arc", input=gis_files[i]+"_t", output=out_dir+"\\"+ascii_out[i])
print "Cleaning up ..."
print ""
### remove internal tmp stuff ###
grass.run_command("g.mremove", rast="*_t", flags="f")
grass.run_command("r.mask", "r")
print "swim2ascii DONE!"
```

Appendix 3.2. SITE output to SWIM input converter

```
"""
Script to load landuse raster provided by SITE into input file for SWIM.
"""
import os, re
import csv
print "Starting Script to include landuse raster from SITE and recreate *.str for SWIM..."
print ""
print "Reading settings..."
print ""
### SETTINGS ###
workingDir=os.getcwd()
# INPUT #
# static -> already in GRASS location
basin = "subbasins"
soil = "soil"
manag = "manag"
wetlands = "wetlands"
mask_swim="mask_swim" # mask with extent used in SWIM
lu_init = "landuse_init" # landuse with extent used in SWIM (slightly larger than extent used in SITE)
#mask_swim = "mask_swim" # mask with extent used in SWIM
hydmask = "hyd_mask" # random hydrotope mask created with 'r.random.cells hyd_mask distance=0'
(unique random number per pixel of study area)
# dynamic for exchange -> load into GRASS
exchange_dir = workingDir+"\dataExchange"
landuse_exchange = exchange_dir+"\landuse_0.asc" #this has to loop according to number of years of
simulation

# SWIM project name (small letters only!!!) -> defines name of output necessary for SWIM
SWIM_PROJECT = "kzn"
# SWIM working directory
SWIM_DIR = workingDir
# OUTPUT #
hyd = "hydrotopes" # output of r.cross and basis for *.str file
hyd_recl = "hyd_recl" # reclassified hydrotope raster (order as defined by *.str file) used to convert
SWIM GIS output into raster
```

```python
# GRASS SETTINGS #00
# initialize grass session
gisbase = os.environ['GISBASE']
gisdb = workingDir+"\grassdata"
location = "South_Africa"
mapset = "PERMANENT"
import grass.script as grass
import grass.script.setup as gsetup
gsetup.init(gisbase, gisdb, location, mapset)
### CALCULATE NEW HYDROTOPES ###
os.chdir(exchange_dir)
print "Importing new land use ..."
print ""
# define mask
grass.run_command("r.mask", "r")
grass.run_command("r.mask", input=mask_swim)
# import landuse map from SITE
grass.run_command("r.in.arc",    input=landuse_exchange,    output="landuse_t",    type="CELL",
overwrite=True)       #grass.run_command("r.in.arc",  input=exchange_dir+"\\"+landuse_exchange,
output="landuse_t", type="CELL", overwrite=True)
# append landuse map - area processed by SITE slightly smaller than area for SWIM
grass.mapcalc("landuse2_t=if(isnull(landuse_t), landuse_init, landuse_t)")
print "Calculating hydrotopes (r.cross) ..."
print ""
# calculate hydrotopes and export stats
grass.run_command("r.cross", input=basin+",landuse2_t," + soil + "," + manag + "," + wetlands + ","
+ hydmask, output=hyd, overwrite=True)
grass.run_command("r.stats", input=hyd, output="hydrotopes_str_t", flags="acln")
print "Creating $SWIM_DIR/${SWIM_PROJECT}.str ..."
print ""
# adjust hydrotopes_str_t and save as $SWIM_PROJECT.str to $SWIM_DIR
# replace ";" and "category" by "" | truncate more than one blank to one blank
with open("hydrotopes_str_t","rb") as source:
    rdr= csv.reader(source, delimiter = ' ')
    with open(SWIM_DIR+"\\"+SWIM_PROJECT+".str","wb") as result:
        wtr= csv.writer(result, delimiter='\t' )
        wtr.writerow(["sub","lu","soil","manag","wet","area","ncell"])
```

```
    for r in rdr:
        #row = [ w.replace(";", "") for w in (r[2], r[4], r[6], r[8], r[10], r[13], r[14]) ]
            row = [ w.replace(";", "") for w in (r[3], r[5], r[7], r[9], r[11], r[13], r[14]) ]
        wtr.writerow(row)
    result.write("0")
```

print "Reclassifying hydrotop raster to \"hyd_recl\" (use this for visualisation of SWIM GIS output) ..."

print ""

```
# reclassification of hydrotopes to match order of SWIM GIS output (ascending order of subbasins)
# create reclassification file
with open("hydrotopes_str_t","rb") as source:
    rdr= csv.reader(source, delimiter = ' ')
    with open("reclass_hyd_t","wb") as result:
        wtr= csv.writer(result, delimiter=' ' )
        for r in rdr:
            wtr.writerow( (r[12],"=", r[0]))
# reclassify
```

grass.run_command("r.reclass",   input=hydmask,   output=hyd_recl,   rules="reclass_hyd_t", overwrite=True)

print "Cleaning up ..."

print ""

```
### remove internal tmp stuff ###
for f in os.listdir(exchange_dir):
    if re.search("_t$", f):
        os.remove(f)
```

grass.run_command("g.mremove", rast="*_t", flags="f")

grass.run_command("r.mask", "r")

print "ascii2swim DONE; back to CONTROL program"

Appendix 3.3. SITE-SWIM coupling script

```python
#!/usr/bin/env python
import os
import sys
import shutil
import subprocess
from subprocess import*
import time
start=time.time()
mainPath=os.getcwd()
swimExe=mainPath+os.sep+"swim"

siteExe=mainPath+os.sep+"SITE"+os.sep+"bin_rel"+os.sep+"SITECmd"
arg='-f drakenfig.ini -steps 3' #make it generic...!
print mainPath
print "Coupled model run on progress..."
pSwim=subprocess.Popen(swimExe,shell=True , stdout=subprocess.PIPE,
stderr=subprocess.STDOUT)
while 1:
    lSwim=pSwim.stdout.readline()
    if not lSwim:
        break
    print lSwim
pSite = subprocess.Popen(siteExe +" " + arg,shell=True , stdout=subprocess.PIPE,
 stderr=subprocess.STDOUT)
while 1:
    lSite=pSite.stdout.readline()
    if not lSite:
        break
    print lSite
pSwim.wait()
pSite.wait()
pSite = subprocess.Popen(siteExe +" " + arg,shell=True , stdout=subprocess.PIPE,
stderr=subprocess.STDOUT)
print pSite.stdout.readline()
while 1:
```

170

```
    lSite=pSite.stdout.readline()
    print lSite
    print lSite
pSwim.terminate()
pSite.wait()
pSite.wait()
while 1:
    i=1
    Substr="Archive: Year"
    line = pSite.stdout.readline()
    if not line:
        break
    elif Substr in line:
        print "Exec SWIM here... !"+ str(i)
        i=i+1
    print line
pSite.wait()

end=time.time()
print "\nCoupled model run completed!"
print "Computation time: "+str(end-start)+" seconds"
raw_input("Press any key to exit")
sys.exit(1)
```

Appendix 4. Land-cover classification on GEE:code snapshot

```
//Land cover training and classification for the Upper Blue Nile
// Load the image to classify
var LS217072 = ee.Image('LANDSAT/LC8_L1T_TOA/LC82170722015268LGN00');
Map.addLayer(LS217072, {bands:['B5','B4','B3']},'2015-01-01');
// Get its geometry
var ROI = ee.Geometry(LS217072.geometry());
Map.centerObject(LS217072, 9);
Map.setCenter(37.6, 10.7, 7);
// Load the training polygons
var trainset = ee.FeatureCollection('ft:1SMcaGSGqV5rRGtEFHqEEM2IR1Lk7BH_NyrhD1eEJ');
Map.addLayer(trainset,{color:'00FF00'},'training');
// Add texture bands
var square = ee.Kernel.square({radius: 70, units:'meters'});
var img2 = LS217072.addBands(LS217072.select(['B3','B5','B6','B7'])
  .multiply(255).byte().entropy(square));
var addpixid = function(image){
  var ID = image.select(['B1'],['ID']).add(image.select('B4'))
    .add(image.select('B2')).add(image.select('B3'));
  return image.addBands(ID);
};
img2=addpixid(img2);
// Train a classifier
var bands = ['B1','B2', 'B3', 'B4', 'B5', 'B6', 'B7', 'B3_1', 'B5_1', 'B6_1','B7_1'];
// use sample() rather than sampleRegions() in order to subsample
var trainsetimg = trainset.reduceToImage(['class'],ee.Reducer.mode());
var img3 = img2.addBands(trainsetimg);
var training = img3.sample({region:ee.Geometry(trainset.geometry()),
  factor:0.7,scale:30});
var trainedRF = ee.Classifier.randomForest(3,0,1,0.5,false,1).train(
  {features:training, classProperty:'mode',
    inputProperties: bands});
var classRF = img2.select(bands).classify(trainedRF);
// Visualize the classification
var classpalette = [
```

```
'8A2BE2','008000','F0E68C','00BFFF','7CFC00','808000','DCDCDC','FF00FF','DC143C',
'4B0082', 'F5F5F5'
];
Map.addLayer(classRF, {min: 1, max: 11, palette: classpalette}, 'classified');
// Clip to a particular geometry
var clip1 = classRF.clip(ROI);
// Select the three classes of interest
var pl = clip1.select('classification').eq(1)
var yp = clip1.select('classification').eq(8)
var cl = clip1.select('classification').eq(9)
// Run the operation at 30-m resolution for compatibility with Landsat pixels
var info = LS217072.getInfo()
var crs = info.bands[0].crs
var crs_transform = info.bands[0].crs_transform
var pl2 = pl.updateMask(pl).connectedPixelCount(1000).gte(500)
pl2 = pl2.updateMask(pl2).reproject(crs, null, 30)
var yp2 = yp.updateMask(yp).connectedPixelCount(1000).gte(500)
yp2 = yp2.updateMask(yp2).reproject(crs, null, 30)
var cl2 = cl.updateMask(cl).connectedPixelCount(1000).gte(500)
cl2 = cl2.updateMask(cl2).reproject(crs, null, 30)
// Reduce the patches to vectors, by class
var polys1 = pl2.addBands(ee.Image.pixelArea()).reduceToVectors({
  reducer: ee.Reducer.mean().setOutputs(['type']),
  scale: 30,
  geometry: ROI,
  maxPixels: 2e11
});
// Include a band with a different value for each type
var polys2 = yp2.addBands(ee.Image.pixelArea().multiply(2)).reduceToVectors({
  reducer: ee.Reducer.mean().setOutputs(['type']),
  scale: 30,
  geometry: ROI,
  maxPixels: 2e11
});
var polys3 = cl2.addBands(ee.Image.pixelArea().multiply(3)).reduceToVectors({
  reducer: ee.Reducer.mean().setOutputs(['type']),
  scale: 30,
```

```
  geometry: ROI,
  maxPixels: 2e11
});
Map.addLayer(polys3,{},'cl');
// Create feature collection
var allpolys1 = polys1.merge(polys2);
var allpolys = allpolys1.merge(polys3);
// Simplifiy the complex geometries
var simpfun = function(feature){
  return feature.simplify(60);
};
var simp1 = polys1.map(simpfun);
var simp2 = polys2.map(simpfun);
var simp3 = polys3.map(simpfun);
//export results as kml
Export.table.toDrive({collection:simp2,
 description:'polys2_217072_6-30',
 fileFormat: 'KML});
```

Samenvatting

Land en water zijn twee van de meest belangrijke, en onderling afhankelijke, natuurlijke hulpbronnen, die essentieel zijn voor het menselijke bestaan. De toenemende wereldbevolking en de wereldwijde economische groei versnellen de vraag naar land voor landbouw en stedelijke ontwikkeling, maar ook de vraag naar water voor irrigatie, energieopwekking en industrialisatie. Het landschap verandert voortdurend als gevolg van deze sociaal-economische factoren. Daarnaast beïnvloeden biofysische factoren, zoals de topografie, klimaatverandering en de variabiliteit van neerslag, het landgebruik en besluiten over veranderingen in landgebruik. Waterbronnen staan eveneens onder grote druk door overmatig gebruik, vervuiling, en veranderingen in de hydrologische processen als gevolg van zowel de sociaal-economische en biofysische factoren. Land en water worden beschouwd als onderling afhankelijke natuurlijke hulpbronnen die elkaar sterk beïnvloeden. Hoewel deze wisselwerking binnen de wetenschap algemeen bekend is, worden land en water meestal beheert onder afzonderlijke management systemen (Le Maitre et al., 2014). Als gevolg hiervan worden veranderingen in landgebruik en de hydrologie gewoonlijk afzonderlijk bestudeerd. Ontwikkelaars van simulatiemodellen nemen deze bestaande scheiding in het beheer van land en water over in hun modellen, wat leidt tot een versimpelde benadering van de ene natuurlijke hulpbron in de modeleerstudies gebaseerd op de andere hulpbron. Het resultaat is dat hydrologie en water vaak worden beschouwd als processen en hulpbronnen die alleen beïnvloed worden door biofysische factoren, terwijl antropogene factoren vaak worden genegeerd ondanks dat deze factoren wel degelijk de hydrologische cyclus kunnen beïnvloeden, bijvoorbeeld door directe effecten op landgebruik.

De noodzaak om te begrijpen, en ruimtelijk de wisselwerking weer te kunnen geven, tussen de hydrologisch cyclus enerzijds en veranderingen in landgebruik anderzijds is absoluut noodzakelijk voor het duurzaam beheren van natuurlijke hulpbronnen in het algemeen, en land en water in het bijzonder. In de afgelopen jaren zijn interessante sub-disciplines zoals 'socio-hydrology' in opkomst, die het belang benadrukken van sociaal-economische factoren en antropogene effecten op de hydrologische cyclus en watervoorraden. De wisselwerking tussen land- en watergebruik is echter nog steeds niet expliciet en dynamisch gekoppeld in de meeste wetenschappelijke simulatiemodellen die beschikbaar zijn voor de analyse van geïntegreerd land en water beheer. Bovendien, vaak als gevolg van de beperkingen in de toegang tot informatie technologie, zijn de kaders die van belang zijn voor de communicatie van de bevindingen van geïntegreerde analyse van land en water niet voldoende beschikbaar om geoperationaliseerd te worden door managers en beleidsmakers. Dit is vooral het geval in ontwikkelingslanden. Voor dit proefschrift is een geïntegreerde evaluatiemodellen aanpak ontwikkeld om het analyseren en modelleren van de wisselwerking tussen land en water met de nadruk op twee geselecteerde rivierbekken in Afrika, de Blauwe Nijl in Ethiopië en de Drakensbergen in Zuid-Afrika. Het onderzoek richt zich op het analyseren van de onderliggende oorzaken van de veranderingen in landgebruik, het evalueren van dynamische terugkoppeling tussen landgebruik en hydrologische modellen, en het ontwikkelen van methoden en instrumenten voor een betere analyse en simulatie door het model. Voor het verbeteren van de toegang tot geïntegreerde analyse en modellering van land en water, zijn verschillende open-source technologieën en standaarden ontwikkeld en toegepast, die de toegang tot gegevens, berekeningen en communicatie van de resultaten makkelijker maken. Uit de resultaten van de studie is gebleken dat in de bestudeerde regio's de grootste veranderingen in landgebruik zijn waargenomen in de afgelopen twee tot drie decennia. De veranderingen hebben geleid tot uitbreiding van landbouwgronden ten kosten van andere landgebruik, zoals bossen en grasland. De belangrijkste drijfveren en factoren die deze veranderingen in landgebruik zijn een toename van de bevolking, de relatief grote afstanden tot wegen en stedelijke gebieden, en topografische factoren zoals de steilheid van berghellingen. Uit de modelresultaten van de wisselwerking tussen de veranderingen in landgebruik en hydrologie blijkt dat dynamische veranderingen in landgebruik de hydrologische cyclus beïnvloedt. Dit is aangetoond door middel van veranderingen in de waterstromen in reactie op de veranderingen in het landgebruik. Deze invloeden zijn duidelijker zichtbaar tijdens de regenseizoenen waarin hoogwater optreedt. Ook wordt aangetoond dat hydrologische processen en de beschikbaarheid van watervoorraden besluiten over de geschiktheid en toewijzing van landgebruik beïnvloeden.

De proefschrift concludeert dat het onderzoek naar de onderliggende oorzaken van sociaal-economische en biofysische veranderingen in landgebruik, en het ruimtelijk expliciete simuleren van de wisselwerking tussen land en water in wetenschappelijke modellen is

noodzakelijk voor een duurzaam beheer van deze natuurlijke hulpbronnen. Dit onderzoek kan bijdragen aan ontwikkeling van geïntegreerde modellen voor het duurzaam beheer van natuurlijke hulpbronnen in het algemeen, en aan land en water beheer in het bijzonder door:

1. Het identificeren van methoden voor het analyseren van onderliggende oorzaken van veranderingen in landgebruik en model parameters,

2. Ontwikkelen van methoden voor het analyseren van de dynamische en ruimtelijk expliciete geschiktheid van landgebruik,

3. Het dynamische kwantificeren van de wisselwerking tussen landgebruik en hydrologische modellen en het analyseren van implicaties van deze interactie voor het duurzaam beheer van ecosystemen in stroomgebieden,

4. Het ontwikkelen van kaders voor een vereenvoudigde toegang tot data, modelberekening en de communicatie van modelresultaten en het faciliteren van het opbouwen van mondiale databestanden voor geïntegreerde simulatiemodellen.

De proefschrift wordt afgesloten met aanbevelingen en suggesties voor verder onderzoek voor de verbetering van methoden, benaderingen en instrumenten die worden gebruikt voor geïntegreerd beheer van land en water. Belangrijkste onderwerpen voor verder onderzoek zijn:

1. Onderzoek naar de haalbaarheid van ingebouwde dynamische modules voor het simuleren van veranderingen in landgebruik in hydrologische modellen, en omgekeerd, om de complexiteit van het simuleren van wisselwerking tussen land en water te vergemakkelijken.

2. Onderzoek naar de toepasbaarheid van de in dit proefschrift toegepaste mondiale databestanden en methoden voor het bepalen van gewas-specifieke geschiktheid van land voor operationele doeleinden en besluitvorming.

ማጠቃለያ

መሬትና ውሃ ጥብቅ መስተጋብር ያላቸውና፤ ለሰው ልጆች ህልውና እና ብልፅግና ጠቃሚ የሆኑ የተፈጥሮ ሃብቶች ናቸው። የህዝብ ብዛት እድገትና ዓለም አቀፍ የምጣኔ ሃብት መስፋፋት፤ የከተማ፤ የመስዋ እርሻ፤ የሃይል ማመንጫ፤ የኢንዱስትሪና የመሳሰሉ ፍላጎቶች፤ በመሬትና በውሃ ላይ ከፍተኛ ጫና አየፈጠሩ ነው። የመሬት ሽፋንም በነኝህ ግፊቶች የተነሳ ተለዋዋጭነት አያጨማሪ ይሆናል። የመሬት ተስማሚነት፤ የአየር ጠባይ ለውጥ፤ የዝናብ መጠንና ወቅት መለዋወጥ፤ እና የመሳሰሉት መልካምድራዊ ግፊቶች የመሬት ሽፋን ለውጥና ተያያዥ ውሳኔዎች ላይ ተጽዕኖ ያሳድራል። የውሃ ሃብቶችም፤ ከማህበራዊ አካባቢያዊ ግፊቶች የተነሳ ብክለት፤ ከአባዛብ በላይ አጠቃቀምና፤ የስነውሃዊ ሂደት ለውጥ ጫናዎች አያጋጠሟቸው ይገኙ።

የመሬት እና የውሃ ሃብቶች ጠንካራ መስተጋባር እና ትስስር አዲስ ሳይንስ ባይሆንም፤ በአብዛኛው ዓለም ሁለቱም በተለዬ አስተዳደራዊ ስርዓት ስር የሚተገበሩ ናቸው። በዚህ የተነሳ፤ የመሬት አጠቃቀም እና የስነሁዊ ሂደት ወይም የውሃ ሃብት በአብዛኛው በተናጠል ይጠናሉ። እነኒህን የተፈጥሮ ሃብቶች ለማጥናት በየጊዜው የሚሰሩ የኮምፒውተር ሞዴሎችም፤ ይሄንኑ የመሬትና የውሃ የተለያዬ አስተዳደራዊ መቀቅር፤ አንዳ በማወረድ ይተገብሩታል። የዚህ አይነቱ አሰራር ውጤት፤ የመሬትና የውሃ ሃብት መስተጋባርን በግልፅ ካለመወከሉ ባሻገር፤ የውሃዊ ሂደት በቁሳዊና በስነህይወታዊ የተፈጥሮ ሂደት ብቻ እንጂ፤ የመሬት ሽፋንን በመለወጥ በቀል የሰው ልጅ በተዘዋወሪ የውሃ ሃብት ላይ የሚያሳድረውን ጫና የሚዘነጉ ይሆናሉ። በመሬትና በውሃ ሃብቶች መካከል ያለውን መስተጋበርና ስነሂደት በትክክል ማወቅ፤ በአንዱ ወይም በሌላው ላይ የሚደረግ ለውጥ ምን ያክል ተያያዥ ጥቅም ሊያመጣ ወይም ምን ያክል ተያያዥ ጉዳት ሊያደርስ እንደሚችል ለማወቅና፤ ዘላቂ ልማት በሚያስገኝ መንገድ እንድ አግባቡ ለመዘጋቶች ይጠቅማል።

ይህ ጥናት፤ በመሬት አጠቃቀምና በውሃ ሃብቶችችን መካከል ያለውን መስተጋባር በተሻለ ግልጽነት ሊያሳዬ የሚችሉ ዘዴዎችንና አካሄዶችን በመመርመር፤ በተቀናጀ መንገድ የተፈጥሮ ሃብቶችችንን ለዘላቂ እድገትና ልማት ስለማዋል ያትታል። በአጠቃላይ፤ የዚህ ጥናት ውጤት እንደሚያሳየው፤ ምርምሩ በተካሄደበተው የአባይና የጡካ (ደቡብ አፍሪካ) ተፋሰሶች ውስጥ፤ ባለፉት ሁለትና ሶስት አስርት አመታት ብቻ ከፍተኛ የመሬት ሽፋንና የመሬት አጠቃቀም ለውጥ መኖሩ ታይቷል። ትልቁ ለውጥ፤ እንደሚገመተው በአመዛኙ ክድን እና ከሳር መሬት ወደ እርሻንት ሲሆን፤ ዋና ዋናዎቹ መንስዔዎች ደገም፤ የሰውና የእንስሳት ብዛት መጨመር፤ የመንገድ፤ የገበያ እና የከምፎ ቅርበት መጨመር፤ እንዲሁም መልካምድራዊ ሁኔታዎች ናቸው። የውሃዊ ሂደትና የመሬት ለውጥ ሂደት ሞዴሎችን በማጣመር የተገኘው ግብረመልስ እንደሚያሳየው፤ የመሬት አጠቃቀም ለውጥ፤ በውሃዊ ሂደት ላይ ተፅዕኖ ያሳድራል። ይህ ተፅዕኖ፤ በተለይም በከፍተኛ ወይም በዝቅተኛ የዝናብ ወቅት

የበለጠ የሚስተዋል ነው። በተያያዘም፣ የአንድ አካባቢ የውሃዊ ሂደት ለውጥ፣ የአካባቢው መሬት አጠቃቀም ውሳኔ ላይ ተፅዕኖ ያሳድራል።

በማጠቃለያ፣ ይህ ጥናት የተቀናጀ የመሬትና የውሃ አስተዳደር፣ ለመሬትና ለውሃ ሃብቶች ጤናማ መስተጋብርና አጠቃቀም፣ ብሎም ለዘላቂ አካባቢያዊ ልማት አይነተኛ መንገድ መሆኑ ይጠቁማል።

ለዚህ አይነት የተቀናጀ የፈጥሮ ሃብት አስተዳደር ደግሞ፣ በውሃዊ ሂደትና በመሬት አጠቃቀም መካከል ያለውን የግብረመልስ ሂደት በምÓደሎች በመታገዝ መፈተሽና በትክክል መገንዘብ ያስፈልጋል። ሆኖም ግን፣ ይህ አይነት አሰራር በዋናነት የተለያዩ ጽንስ ሐሳቦችን የሚዳስስ በመሆኑ፣ ብዙ አካባቢያዊ መረጃና፣ ሞደሎችን የማጣመር እውቀትና ውጤታቸውን በትክክል የመተርጎም አቅም ይጠይቃል።

ይህ ጥናት ለተቀናጀ የተፈጥሮ ሃብቶች አስተዳደር እና በተለይም ለተቀናጀ የመሬትና የውሃ ሃብት

ምክንያቶችን የማጥኛ መንገድ በመለየትና በማበርከት፣

፪. የመሬትን ለተለያዩ ጥቅሞች ተስማሚነት ግልፅ በሆነ ምጣኔ ሀብታዊ፣ መልካምድራዊና ማህበራዊ የመፈተሻ መንገዶችን በመተንተን፣

፫. በውሃና በመሬት አጠቃቀም ስርዓት መካከል የሚኖር ግብረመልስ ያለውን ተጨባጭ ተፅዕኖ በመለካት፣ እንዲሁም፣

፬. የተቀናጀ የመሬት አጠቃቀምና የውሃ አስተዳደርን በመረጃ ቁትነትና ብሎም በማሳለጫነት የሚያግዙ ማፅቀፎችን በመፈተሽ፣ አስተዋፅ ያበረከታል።

180

ABOUT THE AUTHOR

Seleshi Getahun Yalew was born in Dessie, Ethiopia, in 1981. He obtained his Bachelor's degree in 2005 from Addis Ababa University in Information Systems after which he worked as assistant lecturer in the Department of Computer Science at Haramaya University, Ethiopia, until 2008. From 2008-2010, he followed the MSc programme in Water Science and Engineering, specializing in Hydroinformatics and Water Management, at UNESCO-IHE Institute for Water Education in Delft, The Netherlands. He joined the department of Integrated Water Systems and Water Governance at the same Institute in 2011 as a PhD fellow. During his MSc and PhD research, Mr. Yalew has worked on various European Union Framework 7 (EU\FP7) projects including EnviroGRIDS, QUAREHAB and AFROMAISON. Between 2015 and 2017, Mr. Yalew worked as a full time researcher in the Farming Systems and Ecology Group of Wageningen University, The Netherlands. His area of research was on the quantification of synergies and trade-offs of multiple ecosystem services in agricultural landscapes. Mr. Yalew is currently a PostDoc researcher at the Copernicus Institute of Sustainable Development of Utrecht University, The Netherlands. He is also a guest researcher of the Water Systems and Global Change Group of Wageningen University. His current research focuses on climate impacts on the global potential of renewable energy, particularly in the context of integrated assessment modeling.

PUBLICATIONS

Peer-reviewed journal papers:

1. **Yalew , S.G.**, Van Griensven, A., Van der Zaag, P (2016). AgriSuit: A Web-based GIS-MCDA Framework for Agricultural Land-Use Suitability Assessment. Computer and Electronics in Agriculture. 128(October 2016):Pages 1–8.

2. **Yalew , S.G.**, Van Griensven, A., Mul,M.L., Van der Zaag, P (2016). Land suitability analysis for agriculture in the Abbay basin using remote sensing, GIS and AHP techniques. Modeling Earth Systems and Environment, June 2016, 2:101

3. **Yalew, S.G.**, M.L. Mul , E. Teferi , A. van Griensven, d, J. Priess, C. Schweitzer, P. van der Zaag. (2016) Land-Use Change Modelling in the Upper Blue Nile Basin. Environments 2016, 3(3), 21

4. **Yalew, S.**, van Griensven, A., Ray, N., Kokoszkiewicz, L., & Betrie, G. D. (2013). Distributed computation of large scale SWAT models on the Grid. Environmental Modelling & Software. Volume 41, March 2013, Pages 223-230

5. Van Griensven A., Ndomba P., **Yalew S.**, Kilonzo F. (2012). Critical review of SWAT applications in the upper Nile basin countries. Hydrol. Earth Syst. Sci 16:3371-3381.

6. Groot, C.J., **Yalew, S.G.**, Rossing,W.A.H., (2018). Exploring ecosystem services trade-offs in agricultural landscapes with a multi-objective programming approach. Landscape and Urban Planning: https://doi.org/10.1016/j.landurbplan.2017.12.008

7. **Yalew , S.**, T. Pilz, C. Schweitzer, S. Liersch, J. van der Kwast, A. van Griensven, M.L. Mul, P. van der Zaag. (2017) Dynamic Feedback between Coupled Land-use and Hydrologic Models. Environmental Modeling & Software (*under review*).

8. **Yalew, S.G.**, Rossing,W.A.H., Bianchi, F., (2017). Trade-off and synergy between ecosystem services: Exploring the potential of semi-natural habitats for maximizing benefits in agricultural landscapes. Ecological Modeling (*under review*).

9. **Yalew, S.G.**, van Griensven, A., van der Zaag, P, (2018). Hydrologic impacts of semi-dynamic land-use change in the Blue Nile basin. Environment Systems and Decisions (under review).

Conference proceedings

10. **Yalew, S.**, Pilz T., Schweitzer C., Liersch S., van der Kwast J., Mul M.L., van Griensven A., van der Zaag P. (2014) Dynamic Feedback between Land-Use and Hydrology for Ecosystem Services Assessment, in: D. P. Ames, et al. (Eds.), International Environmental Modelling and Software Society (iEMSs), San Diego, USA.

11. Mugwizaa, L., **Yalew, S.**, van der Kwast, J., Hamdarda, M., van Deursenb, W., (2014). A Spatial Planning Tool for the Evaluation of the Effect of Hydrological and Land-use Changes on Ecosystem Services, In: Ames, D.P., Quinn, N.W.T., Rizzoli, A.E. (Eds.), International Environmental Modelling and Software Society (iEMSs): San Diego, USA

12. **Yalew, S.**, P. van der Zaag, Mul, M., Uhlenbrook, S., Teferi, E., van Griensven, A., van der Kwast, J. (2013). Coupled hydrologic and land use change models for decision making on land and water resources in the Upper Blue Nile basin, Abstract, EGU General Assembly 2013.

13. Van der Kwast, J., **Yalew, S.**, C. Dickens, L. Quayle, J. Reinhardt, S. Liersch, M. Mul, M. Hamdard, W. Douven, (2013). A Framework for Coupling Land Use and Hydrological Modelling for Management of Ecosystem Services, ICWRE-2013, Geneva.

14. **Yalew, S.**, Teferi, E., van Griensven, A., Uhlenbrook, S., Mul, M., van der Kwast, J., & van der Zaag, P. (2012). Land Use Change and Suitability Assessment in the Upper Blue Nile Basin under Water Resources and Socio-economic Constraints: A Drive towards a Decision Support System, IEMSs2012, Leipzig.

15. **Yalew, S.**, van Griensven, A., Kokoszkiewicz, L. (2010). Parallel computing of a large scale spatially distributed model using the Soil and Water Assessment Tool (SWAT), IEMSs2010, Ottawa.

16. Haest, P.J., Broekx, S., **Yalew, S.**, Boucard P., van der Kwast, J., Seuntjens, P. (2013). A scenario analysis of measures tackling nitrogen discharge in the Scheldt river basin, Proceedings of the 2nd European Symposium, November 2013, Leuven, Belgium.

Conference/workshop presentations and abstracts

17. **Yalew, S.** & van Griensven, A., (2013). The Use of Google's Earth Engine to Assess Loss of Vegetation Cover in the Anger Valley, Upper Blue Nile, Ethiopia, Abstract, OpenWater2013, Brussels.

18. **Yalew, S.**, van Griensvem, A., van der Kwast, J., Hamdard, M., van der Zaag, P. (2014). Challenges and Opportunities of Integrating Models to Support Decisions in Natural Resources Management, AFROMAISON International Symposium and Workshops, May 2014, Delft, the Netherlands.

19. Van der Kwast J., **Yalew S.**, Broekx S., Haest P.J., Seuntjens P., Jesenska S., Carpentier C., Blaha L., Boucard J., Slobodník J., Bastiaens L., van Griensven, A., (2013). REACHER: A Decision Support Tool to Evaluate Scenarios of Measures for the Reduction of Pollution fluxes, Abstract, OpenWater2013, Brussels, Belgium.

20. Xuan Y., **Yalew S.**, Zhu X., Xu Z., van Griensven A. (2011). Automated analysis of upstream- downstream relationships using Bayesian Belief Networks from spatially distributed SWAT models, OpenWater symposium. pp. 68, Delft

21. Seuntjens P., Haest P. J., Broekx S., Van der Kwast J., **Yalew S.**, Hoang L., Van Griensven A.,Boucard P., Blaha L., Jesenska S., Slobodnik J., Bastiaens L., The integration of watershed fate models, ecological assessment and economic analysis of water rehabilitation measures in the REACHER decision support system-Aquarehab WP6, 2nd Water Technology and Management Symposium, November 2013, Leuven, Belgium.

22. **Yalew, S.**, Van Griensven, A., and Van der Zaag, P. (2015). The Use of Google Earth Engine for Land Cover Classification in the Upper Blue Nile Basin. OpenWater2015 Symposium, Addis Ababa, Ethiopia.

23. **Yalew, S.**, Rossing,W., Bianchi, F., Groot, J., and Fleskens, L. (2015). Modelling Trade-offs between Erosion Prevention and Other Ecosystem Services using the Landscape IMAGES Framework. OpenWater2015 Symposium, Addis Ababa, Ethiopia.

24. Geertsema, W., **Yalew,S.**, Rossing, W., van der Werf, W., Bianchi, F. (2016). The role of costs and benefits in optimizing agricultural landscapes for multiple ecosystem services. European Ecosystem Services Conference, Antwerp, Belgium.